IMPACT OF IRRIGATION ON AGRICULTURAL PRODUCTION

IMPACT OF IRRIGATION ON AGRICULTURAL PRODUCTION

N.T. SOMASHEKAR
M.A., M Phil., LL.B.
Lecturer, DOS in Economics,
Government Arts College,
Hassan, Karnataka

Dr. S. MAHENDRA KUMAR
M.A. Ph.D.
Reader, DOS in Economics and Co-operation,
University of Mysore,
Manasagangothri, Mysore, Karnataka

ANMOL PUBLICATIONS PVT. LTD.
NEW DELHI - 110 002 (INDIA)

ANMOL PUBLICATIONS PVT. LTD.
H.O.: 4374/4B, Ansari Road, Daryaganj,
New Delhi-110002 (India)
Ph.: 23278000, 23261597
B.O.: No. 1015, Ist Main Road, BSK IIIrd Stage
IIIrd Phase, IIIrd Block,
Bangalore - 560 085 (India)
Visit us at: www.anmolpublications.com

Impact of Irrigation on Agricultural Production

First Edition, 2009

ISBN 978-81-261-3812-8

PRINTED IN INDIA

Printed at Mehra Offset Press, Delhi.

Dedicated

to

The Lotus Feet

of

Sri Lakshmi Venkateshwara Swamy

CONTENTS

PREFACE

Irrigation in modern times has been adopted as technology in a sense that application of water to the growth of different crops suiting to the soil features and crop requirements has been timely followed in a scientific manner. Hence, the irrigation is an innovative break-through in the traditional type of agricultural practices and methods. As a result, a number of changes take place accounting for different types of benefits. The present book has made an attempt to introduce to the reader all the aspects of minor irrigation—its assessment, development, utilization and management with special reference to tube well irrigation in lucid style.

This book is intenaed as a reference book for post-graduate students in economics, agriculture and engineering. It will be a valuable reference to research scholars, agricultural scientists and to all field workers in the area of irrigation water management. It will also be useful to the bankers, farmers, estate owners and others dealing with tube well irrigation systems.

In the preparation of this book, I have received immense help from a number of persons. In the first place, it is my duty to express my regards and gratitude to my guide Dr. S. Mahendrakumar M.A., Ph. D., Reader, DOS in Economics and Co-operation, University of

Mysore, Manasagangothri, Mysore for his valuable guidance, kind suggestions and for the invaluable time devoted to discussions and corrections involved while preparing this volume.

I am highly thankful to Dr. B. Subbulakshmi, Professor and Head, Department of Economics, Directorate of Distance Education, Madurai Kamaraj University, Madurai for her valuable suggestions and permitting me to undertake research work on approved topic. I am grateful to Dr. M. Jayaprakash M.A., Ph.D., Post Doctoral Fellow, DOS in Economics & Co-operation, University of Mysore, Manasagangothri, Mysore for his help in every respect through out the processing of the present work.

My sincere thanks are due to our Joint Director Sri Hari, Collegiate Education, Regional Office Mysore and Prof. T. N. Prabhakar, Principal and Director, P. G. Centre, Government Arts College, Hassan for providing this opportunity by permitting me to undertake the research work.

Several colleagues and friends of mine have helped me in innumerable ways in this endeavour. Specially, I should mention the names of Dr. Gopalsingh, coordinator, (Dos in Economics, Govt. Arts College Hassan), Dr. I. A. Mohan Principal, Govt. Womens College, Hassan, Dr. Rajanna, Prof. H. K. Lalitha Devi, Prof. K.T. Krishnegowda, Prof. H. S. Ravindra, Mr. R. Radhakrishna, A. Veeranna, Mr. M. B. Irshad, Mr. K. R. Narendra B. E., Mr. Srinivasa Murthy Anmol Publications, New Delhi, Mr. A. N. Chandrappa (C. R. Pattna), and college staff. I am very grateful to all of them.

My greatest debt is to my parents, my wife Smt. Sujatha and my son N. S. Swaroop who made many personal sacrifices to enable me to complete this work. I also express my grateful thanks to the Anmol Publications Pvt. Ltd., New Delhi for bringing out this book in a record time.

Constructive criticisms are always welcome, appreciated and the same will be incorporated in the next edition.

N. T. SOMASHEKAR
S. MAHENDRA KUMAR

1

INTRODUCTION

Farmers cannot always depend solely on rainfall to sustain their crops. In fact, many of the world's agricultural regions receive little or no precipitation. In these areas, water must be diverted from streams or lakes or drawn from wells. The result has been the development of irrigation systems. Irrigation is the method of supplying water to land to support plant growth. This technology has had a powerful role in the history of civilization. Mesopotamia, is considered as the cradle of civilization, had one of the earliest water diversion systems. Agriculture began there about ten thousand years ago. Water channels helped divert water from the Tigris and Euphrates Rivers during the dry months and also helped control the annual July floods. The civilization that developed in Egypt owed its existence to the water diversion schemes that delivered water to the fields from the Nile River and its delta. The ancient Egyptians invented the *shaduf*, a long pole with a bucket suspended at one end and a counterweight mounted at the other. The pole pivoted vertically to lift water in the bucket to a higher level. *Shadufs* are still in use. About 2000 B.C., the Chinese were diverting water from the Hwang Ho and Yangtze Rivers and building levees to keep their rice *paddies* flooded. In approximately 130 A.D. Native Americans built canals, terraces, and check dams that are still visible today in the Sonoran Desert of Mexico and Arizona.[1]

The Roman Empire drew much of its stability from public works projects that included canals and *aqueducts*. The Romans had several methods of lifting water, including the treadmill-powered chain of buckets, the human-powered scoop wheel, the water-driven tympanum, and the Archimedean screw. Today, mechanical pumps raise water, while channels and sluice gates regulate its level and flow. Portable pumps are available to farmers who wish to irrigate on a short-term basis, as in a drought, and large permanent lift stations operate in larger irrigation schemes.

Irrigation networks have been constructed in many parts of the world, such as in the rice fields of eastern Arkansas, the olive groves of southern Spain, and the wheat fields of Victoria and Australia. One of the most ambitious irrigation efforts has taken place in the Sacramento, San Joaquin, and Imperial Valleys of California. An important figure in the agricultural development of California was Harriet W. R. Strong. Left to fend for herself after her husband's suicide in 1883, she acquired a number of patents for irrigation and flood control measures, including, in 1887, the construction of a series of dams so that water backed up from one helped support the dam immediately upstream from it. Irrigation methods also include the use of sprinkler systems. Field-length pipes supported by wheeled legs water of the cornfields of the Midwestern United States. Similarly, the potato fields of Idaho and Oregon are watered by mobile pipes that revolve in a circle. Irrigation has its environmental drawbacks, causing leaching in some areas. Water tables have been depleted from over-pumping, as in the cotton fields of west Texas. Water diversion also has fostered regional disputes over water rights, as is the case among the states of the upper and lower Missouri River Valley. In spite of its financial and environmental costs, irrigation has been and will be a vital part of man's existence for the foreseeable future.[2]

Indian economy is agricultural in the sense that at present about 22 per cent of National Income is contributed by the agricultural sector and nearly 59 per cent population earns its livelihood from agriculture. Along with this about 26.10 per cent population lives below poverty line. Backward agriculture and increase in population at the rapid rate of 1.27 per cent per annum have created a severe problem of acute food shortage. The result is underfed population, under nutritional food and the mass poverty, inadequacy of food necessitated imports of food grains from the period of the Second Five Year Plan till a few years in the recent past. This exerted a heavy pressure on foreign exchange resources and to that extent retarded economic development.

Self-sufficiency in food, creation of marketable surplus so as to promote agro-based industrial development and also to earn foreign exchange through export of food items demand agricultural development at an accelerated rate. Removal of agricultural backwardness so as to increase agricultural productivity is the only solution.

Agricultural development mainly depends on two basic pre-requisites; soil and water. Deep black soil which holds moisture for a period long enough to promote plant growth is limited. The natural source of water for agriculture is the rain. Indian agriculture depends largely on vagaries of monsoon, which is uncertain, irregular, and unevenly distributed throughout the year and over an area. In heavy rainfall areas the rivers drain out the water and hence it is wasted for want of storing and preserving devices. In addition, it creates problems of soil erosion and flood damage. In scanty rainfall areas, however, there is an acute shortage of water for agriculture.

Water shortage for agriculture is also the main cause of rural poverty. Rural uplift is a focal point of our Plan strategy. Water wastage in heavy rainfall areas, and water shortages elsewhere draw our attention to the urgent need of creating proper and adequate irrigation facilities for overall agricultural development. The Planning Commission has, therefore, attached a marked significance to the problem of irrigation right from the First Five Year Plan. Mr. Morarji Desai, as the Chief Minister of the Bombay State, emphasized this point in a radio talk on January 14, 1953. He said, "I have no doubt the (Planning) Commission has indicated a right order of priorities by emphasizing the development of agriculture, particularly irrigation, with a view to securing self-sufficiency in food and development of power and certain basic industries for our industrial expansion".[3]

Agriculture needs assured and regular supply of water cannot be left to the vagaries of monsoon. Dependable source of water can be had from irrigation facilities alone. Irrigation implies a deliberate human effort to carry water off the natural water sources to the crops in fields. Irrigation brings about a drastic change in the soil. This facilitates the use of manure which gives high yield even in comparatively poor soils. Intensive use of lands, double and treble cropping, and introduction of cash crops bring not only stability to agriculture but also create surplus in the economy. It generates more income to rural population and paves the way to economic prosperity. "Thus irrigation is not only important for stabilization of agriculture, but is the *sine qua non* for all-round economic development."

Therefore, the Union and all the State Governments are attempting to make optimum use of irrigation potential and sources of irrigation.

IMPORTANCE OF IRRIGATION

Water resources are important wealth and prosperity of any country largely depends on the efficient use of water resources for agricultural production and electricity generation. Crop production can be increased many folds (4 to 10 times) when irrigation is provided in semi-arid tropics where rainfall is inadequate, erratic, ill distributed and often leads to drought conditions. Therefore, future crop production under irrigation conditions depends solely on efficient and judicious use of water to realize the cherished gains from irrigation.

Irrigation has proved beneficial to the country. In fact, "Irrigation forms the datum line for sustained successful agriculture." It alleviates suffering, preserves life, averts famines and advances the material prosperity of the country. In fact, as pointed out by Sir Charles Trevelyan, "Irrigation is every thing in India; water is more valuable than land, because, when water is applied to land it increases its productiveness at least six fold and renders great extents of land productivity, which otherwise would produce nothing or next to nothing."[4] Dr. Knowles writes, "The irrigation works have increased the yields, and the value of the land and the revenue derived from it. They have lessened the cost of famine relief and have helped to civilize the whole region. In addition, they yield handsome profits to the government."[5]

An irrigation project confers a number of benefits, both direct and indirect on the nation. By direct benefits we mean increases in all farm production as a result of provision of irrigation facility. In addition to direct benefits, irrigation may also give indirect benefits to the region irrigated by creating opportunities for processing activities as a result of increased agricultural productivity. It may create certain newer activities, providing more investment, employment, income and other benefits. The greatest benefit of irrigation, however, is the sense of security engendered in the minds of the farmers and stability of agricultural business, which an irrigation system confers on the area irrigated. The direct benefits of irrigated area, by and large, is calculable in quantitative terms. But it is not easy to do so in case of indirect benefits because development of trading and industrial activities in the area might have simultaneously been influenced partially by other economic forces.

The following reasons explain the importance of irrigation in agriculture.

1. Insufficient, Uncertain and Irregular Rains

The period of rainfall is restricted to only four months in a year, June to September, when monsoon arrives. The remaining eight months are dry. There is some rainfall during the months of December and January in some parts of the country. Even during monsoon, the rainfall is scanty and undependable in many parts of the country. Sometimes the monsoons are delayed considerably while sometimes they cease prematurely. This pushes large areas of the country into drought conditions. Proper development of irrigation facilities can help the country in solving the problems created by insufficient, uncertain and irregular rains. With the help of irrigation, droughts and famines can be effectively controlled.

2. Higher Productivity on Irrigated Land

Productivity on irrigated land is considerably more than the productivity on unirrigated land. Dr. Pandse and his associates have shown that use of irrigation can help to produce 0.5 to 0.7 tonne more of food grains per acre. The Planning Commission has estimated the productivity of irrigated land to be almost double (93 per cent more) compared to that of unirrigated land.[6] B.D. Dhawan has observed that 22 quintals crop per hectare in 1983-84 whereas it was less than 9 quintals crop per hectare on unirrigated lands. He further argues that the productivity differential five years later would be higher than 13.3 quintals observed during 1983-84.[7] This is because of a firm upward tendency in the overall irrigation yield that has been witnessed during the 14-year period 1970-71 to 1983-84. In a later study, Dhawan showed that the productivity on irrigated plots in 1992-93 tended to be 2.3 times that on unirrigated plots.[8] This shows that production and productivity in agriculture can be increased considerably by extending irrigation facilities.

3. Multiple Cropping Possible

Since India has a tropical and sub-tropical climate, it has potentialities to grow crops on a year round basis. However, since 80 per cent of the annual rainfall is received in less than four months, multiple cropping is generally not possible. Provision of irrigation facilities can make possible the growing of two or three crops in a year in most areas of the country. This will considerably enhance agricultural production and productivity.

4. Role in New Agricultural Strategy

The successful implementation of the High Yielding Varieties Programme depends, to a large extent, on the timely availability of ample water supply. The high-yielding varieties of seeds and chemical fertilizers require substantial water at regular intervals of time. This is the reason why the HYVP has remained limited to irrigated areas only.[9] Therefore, benefits of new strategy can be extended only if more irrigation facilities are made available to larger areas of land.

5. Bringing more Land under Cultivation

The total reporting area for land utilization statistics was 306.06 million hectares in 2003 in India. Of this, 19.25 million hectares were barren and uncultivable land, 11.68 million hectares fallow land other than current fallows lands. Current fallows lands include, lands which are lying fallow for less than one year, other than current fallows lands includes lands lying unploughed for one to five years. Culturable waste land comprises another 13.49 million hectares. Cultivation on all such lands is impossible in some cases while in others it requires substantial capital investment to make land fit for cultivation. Even then, the possibility cannot be ignored that provision of irrigation facilities can make some portion of this land cultivable.

6. Reduces Instability in Output Levels

A study of eleven States during the period 1971 to 1984 has shown that irrigation helps in stabilizing the output and yield levels. More specifically, irrigation has lowered the standard deviation in annual growth rates from 19.0 to 7.3 per cent in the case of output of all crops, from 14.5 to 5.9 per cent in respect of overall crop yield, and from 5.0 to 2.4 per cent in respect of overall crop acreage.[10] Not only this, irrigation plays a 'protective' role during drought years. Since both income and employment are positively and closely related to output, prevention of fall in output during drought is an important instrument for achieving stability of income and employment in the countryside. Irrigation has enabled many States to acquire 'partial immunity' from drought. The vital food grains sector is now substantially insulated against a serious rainfall failure. This comparative insulation "has provided strength and stability to the food grains distribution system of the public sector. For, both procured wheat and rice that constitute the backbone of this system now mainly originate from irrigated areas."[11]

7. Indirect Benefits of Irrigation

Irrigation confers indirect benefits through increased agricultural production. Employment potential of irrigated lands increases, increased production helps in developing allied activities, means of water transport are improved, income of government from agriculture increases, etc. Availability of regular water supply will increase the income of farmers imparting a sense of security and stability in agriculture.

8. Secondary and Tertiary Effect

Increased production by way of irrigation required more infrastructures for marketing and more production of farm inputs. Thus agro-industries can develop at faster rate.

Development of irrigation has thus a very important role in the growth of Indian agriculture.

ROLE OF IRRIGATION

Irrigation is the obvious means of making the country's agriculture relatively independent of the vagaries of rains. It promotes agricultural *vis-à-vis* overall economic development and raise the standard of living of a long established agricultural population. Economic and social development to a great extent depends upon the creation of surplus agricultural produce; this often requires extension of agriculture through irrigation projects and practices to ensure optimum land utilization.

In study area, agriculture mainly depends on monsoons, which are mostly uncertain. The natural supply of water through rainfall is inadequate and unevenly distributed geographically and also over time, and irrigation is the only resort. Major portion of land remains thirsty for want of irrigation facilities. In view of the importance of water for agriculture, emphasis on irrigation has become an issue of inescapable importance in study area. The role of Irrigation is as follows:

- Irrigation provides production against vagaries of monsoons i.e. both famine and floods.
- Irrigation increases the optimum use of land and consequently, it increases agricultural production and promotes better cropping pattern. And also it generates farm income by increasing the productivity of land.
- Irrigation is important for increasing of the production of commercial crops and for increasing the employment in agriculture as well as in industries.
- Irrigation helps the rural families, to cross the poverty line and protects the cultivator from risk

and uncertainty of crop harvest during the drought year.

- It adopts modern inputs like HYV seeds, chemical fertilizers, pesticides, implements, etc.
- It promotes consumption expenditure and investment expenditure of the farmers. And also it improves education, health and hygienic conditions of farmers.
- It changes the attitude of farmers. It is also conducive to other effects such as town planning, tourism development, etc.

The development of irrigation thus has a very important role in the growth of agriculture.

IMPACT ON AGRICULTURE

The development of irrigation has a favourable impact on agriculture in several ways.

- Irrigation makes the outcome of the crop more certain and reduces instability in agriculture and leads to change in the crop pattern towards lucrative and remunerative crops.
- High yielding crops are more biased to adequate irrigation input and there is a good response in the yields of crops with higher doses of fertilizers and pesticides.
- Less risk and certainty of harvest of crops with assured irrigation permit the cultivator to invest more on the inputs which have its own effect on the output and stability in crop production.
- Irrigation by bringing changes in the cropping pattern, by increasing the intensity of cropping

contributes for the increase in the employment in crop production.

- Irrigation is the pre-requisite for the adoption of a new technology in agriculture.
- Investment in irrigation leads to the conversion of dry land into wetland. And irrigation also provides security against the recurrence of droughts.

In view of the importance of irrigation and the fact that water is a scarce resource. It should be utilized economically and efficiently.

WEAKNESS OF IRRIGATION

The following are some of the principal defects of our irrigation system.

- There are wide disparities between districts as far as irrigation is concerned.
- More than half of the irrigated area depends on shallow wells, minor tanks and small diversion works which themselves depend on rainfall.
- Proper drainage facilities are lacking; hence water logging.
- Extensive canal system has been constructed with inadequate water supplies.
- Utilization has been lagging behind the creation of the irrigation potential.
- Delay in the completion of major irrigation projects.
- Lack of well-investigated projects in the pipeline.
- Inadequacy of finance and organization.
- The problem of underutilization of the irrigation.
- Lack of coordination between irrigation and agricultural development.

- Inter State water disputes.
- Increasing costs of irrigation.

NEED FOR MINOR IRRIGATION

Both major and minor irrigation works have relative advantages of their own. The major schemes utilize surplus river waters, which would otherwise run to waste; they benefit large areas, give sure protection in years of scarcity and can often be designed for multiple uses. As against these, minor schemes require comparatively small outlay, yield quick results and can be executed speedily with local resources.

In view of the urgency of our food problem and the inadequacy of resources, especially of foreign exchange, greater attention began to be paid to minor irrigation works. In community development project areas and national extension service blocks, special emphasis is placed on minor irrigation works, since these call forth self-help and popular enthusiasm among agriculturists.

The following are some of the causes of need for minor irrigation system.

- Major irrigation works involve huge investment and take a long period for completion. But minor irrigation requires small investment and they can be completed with in a short period.
- If quick returns are required from investment on irrigation, minor irrigation works are more suitable. Minor irrigation also provides protection against vagaries of monsoons i.e., both famine and floods.
- It absorbs modern inputs like HYV seeds, chemical fertilizers, pesticides, implements, etc. and it also

enhances employment opportunities in the primary and other sectors.

- It raises gross income of farmers and improves health and hygienic conditions of farmers.
- Minor irrigation is essential in order to protect various regions from famines and to ensure double and if possible, triple cropping and rapid increase in agricultural production.

We may now sum up the minor irrigation as definitely a superior method of irrigation system. However, for its successful application certain conditions must be fulfilled.

REVIEW OF LITERATURE

Realizing the importance of minor irrigation, government, research institutions, banks, individual researchers, etc. have undertaken several evaluation studies at the district as well as at state levels. Such studies focused, by and large, on the feasibility of the minor irrigation scheme, cost factor, benefit-cost ratio, employment generation, rate of return on capital, cropping intensity, cropping pattern, etc. A review of available studies is presented below.

The earliest study on irrigation aspect was that of the Famine Commission constituted at the fag end of the 19th century. The Commission, on studying the problem of the then Madras province, felt that the development of minor irrigation works could be more relevant for the province. It also observed that the financial results of expenditure on minor irrigation works had been satisfactory. The total net revenue for 1895-96 and 1896-97 yielded a return of 8.24 and 9.13 per cent respectively on the recorded capital outlay.[12]

A synoptic view of irrigation and agriculture of the world is given by Cantor (1967). His book includes a systematic history of irrigation methods and problems of irrigated agriculture. An year-book (1955) on agriculture of the USA, namely, *Water* aims at explaining the nature, behaviour and conservation of water in agriculture. It also includes some of the broad problems, facts and basic principles of it. Clark (1970) deals with economics of irrigation. Other works carried out by some of the foreign scholars on the method of irrigation and water management are Manneir (1974), Short (1974), Bengtson (1974), Andreae (1975), Garbrecht (1978), Chambers (1988) and Gooneratne (1990).

Irrigation has long been the concern of many a research. In the years after independence, several new schemes have been implemented and several studies followed, on the nature and implications of such irrigation works. Gadgil (1948), in an outstanding research of his, evaluates economic effects of irrigation with reference to the Godavari and the Pravara canals. His study throws light on many direct and indirect benefits of irrigation. Further, he presents the beneficial impact of irrigation in dry areas which are deprived of this facility because of lack or water resources. The study concludes that the provision of irrigation facility to the people of the Godavari and Pravara canal command area has enabled them to have a superior cropping pattern, higher productivity and gross farm income. The indirect benefits have also accrued to the people who have been involved in the processing and transporting of enhanced agricultural production in the command area. The volume of trade has also increased on account of increased agricultural production and it has benefited railway traffic, road traffic and trading establishments, in Ahmednagar district. Irrigated agriculture being labour-intensive has increased the demand for hired labour, which in turn has raised the wage rate. This has improved the economic conditions of labourers also. The construction and maintenance of a project has far-reaching effects on the economic life of the community living within the region and also to some extent on the neighbouring communities as well.

In 1958, the Planning Commission of the Government of India constituted a study group under the chairmanship of D.R. Gadgil to measure the direct (primary) and to the extent possible the indirect (secondary) benefits arising from the selected schemes like the Sarola Canal (Uttar Pradesh), Tribeni Canal (Bihar), Damodar Canal (West Bengal), Cauvery Mettur Project (Tamil Nadu), Nijamsagar Project (Andra Pradesh) and Ganga Canal (Rajasthan).

The studies so commissioned were the *ex-post* assessment of benefits of irrigation provided by different projects with the help of the primary data. The studies have concluded that irrigation leads to the intensive use of land, to shifts from food to commercial crops, increase in cropping intensity, promotion of the use of modern and traditional inputs and increase in gross production and income.

The direct and indirect impacts of irrigation has been analysed in depth by Jha (1967) taking the case of the Tribeni Canal in Champaran district of Bihar. This study is special in that the author makes a comparison between the project area and a control area. The project area was 96 per cent irrigated while the control area was only 34 per cent irrigated. The variables chosen for analysis were land ownership, land use pattern, cropping pattern, rural credit borrowings, input-output rates in the category of direct impact and farm labour, rural industries, rural transport and urban industry in the category of indirect effect. The study has concluded that irrigation has definitely brought prosperity to the cultivators of the project area, as compared with the cultivators of the control area.

A historical analysis of the changes in agricultural landscape in Varanasi district during 1911-1963 suggests that one of the main factors for change is the spread of modern irrigational facilities, especially, canals and tube wells, since independence. The study has shown a sharp fall in the area of dry crops and an increase in paddy and double cropping. It is further suggested that the early paddy is more important in the tube well irrigated areas while late paddy is important in the canal-irrigated areas (Mukherjee, Singh and Mukherjee, 1967).

The study made by Basu and Mukerjee (1963) and ARDC (1980) analysed the income aspect of canals and tube wells respectively. The latter pointed out that the net income for the project areas formed Rs. 1,056 in Bhojpur district, Bihar which was higher than non-project area (Rs. 266). However, these studies have not dealt with aspects such as impact of irrigation on cropping intensity, employment, generation, wages, livestock income etc., etc.

The Irrigation Commission (1972) has pointed out that "the canal water rates in force for the two major irrigated crops, rice and wheat, range between 1.1 and 4.5 per cent of the value of the produce, which is very low.

A large part of the irrigated area in the country receives irrigation from privately owned sources and the farmers concerned incur heavy expenditure on irrigating their crops. Some of them purchase water from owners of neighboring tube wells or lift pumps at a price several times higher than canal rates and some even share a third of their produce in return for water used. It is thus evident that the farmer, realizing the key role of irrigation amongst various inputs, is prepared to pay high water charges and has the capacity to do so." So the commission is of the opinion that, "there is hardly any justification for keeping the rates so low as to cause a loss to the exchequer and a burden on the rest of the community. The low rates which are in force require early upward revision."

A study was conducted at the Uttrakhand Agricultural University, Pant Nagar farms and a dozen similar farms of a few progressive farmers of the Tarai area, during 1966-67 and 1967-68 agricultural seasons. A similar set of data was also collected on half a dozen farms of Madhya Pradesh during 1965-66 agricultural year. These data highlight the percentage contribution of five different modern inputs. Irrigation adds to about 22 per cent towards the total additional benefit. As such irrigation water rates may have to be fixed at something less than 22 per cent, there by enabling the cultivator to retain the difference between the additional net benefits and water rates so fixed as net returns. The difference so arrived at, will form part of the net profit accrued as a result of the use of irrigation water in crop production. This difference should be sufficiently higher to provide incentive to the farmers for maximum utilization of irrigation potential.

The Sixth Finance Commission drew the attention of the State Governments regarding the sustained heavy losses on major and medium irrigation works. As against a marginal loss of about half a crore in 1950-51 it increased to 150 crores in 1971-72 on all the irrigation projects in India. So there is an emphasis on the financial aspects of the irrigation projects. But there are many, who are against this view.

In an article entitled "An Evaluation of Commercial Irrigation Projects in Andhra Pradesh" in *Lok Udyog* of

February 1977. Dr. S. Kishan Rao is of the view that, "Independent India, committed for planned development on the lines of socialistic pattern of society, should view the irrigation projects unmindful of the return, but from overall macro consideration such an increasing national income and self-sufficiency in food grains. Irrigation should not have any productivity test like education and community development, rather should be given a liberal treatment. Estimate of returns should be considered as an economic exercise and should only guide in fixing inters priorities in sanctioning projects."

The distribution of irrigation benefit between the large and small farmers had always been in the favour of large farmers who have the irrigation bureaucracy under its control. Small farmers get less benefit. Robert Wade has stated this fact in an article entitled "Administration and the Distribution of Irrigation Benefits." The author has quoted from D. Thorner's article; "The Week and the Strong on the Darda Canal" and E.J. Valde's, Ph.D. thesis; "The Distribution of Irrigation Benefits; A Study in Haryana" to make this observation stronger.

In terms of Financial Rate of Return (FRR) as well as the benefits from the project investment on minor irrigation sources, a few studies have been made. For instance, the ARDC study in Kota district, Rajasthan (1980), the NABARD study in Pune district (1982), the State Bank of India study in Bilaspur block of Ambala district, Haryana (1985) and the study by the Agro Economic Research Center (AERC) of the University of Delhi in Bhiwani district, Haryana (1985)[13] have attempted on such lines. The studies revealed that the Financial Rate of Return worked out at 26 per cent for all the farmers in Pune district. It was, for the successful wells, 27 per cent as against 17 per cent for incomplete wells in Kota district, 12 per cent for tube wells as against the interest rate of 10 to 25 per cent in Bilaspur block of Ambala district and the FRR of pump sets and sprinklers was over 45 per cent for small farmers in Bhiwani district. These studies too have not covered the aspects like income and employment generation, changes in cropping pattern, cost aspect, etc.

In 1980, the Agricultural Refinance and Development Corporation under took an *Ex-post* Evaluation on Minor Irrigation Schemes in Bhojpur district of Bihar. The object of the study included quantifying the actual benefits and costs to scheme beneficiaries and estimating the benefits to persons who purchase water from the scheme tube wells for irrigation. The finding was that the non-beneficiary got an income of Rs. 679 while the beneficiaries' income accounted for Rs. 2,142. The beneficiaries here referred to are those who purchased water from tube well schemes. As this study has stressed the gross value produce benefits, it has not covered the aspect of employment generation.[14]

In 1984, a study was made in Orissa on the public tube wells and river lifts by NABARD.[15] The study brought to light the fact that the economic rate of return on capital invested in tube wells four per cent and the economic rate of return on river lift irrigation units was negative.

The study by Gopal Swamy and Nagadevara (1984) evaluated minor irrigation schemes in Bangalore district. Its findings revealed that minor irrigation schemes created more employment. The additional employment generated varied between 159 per cent and 452 per cent. Regarding income generation, irrigated areas had recorded 372 per cent increase in income per hectare more than unirrigated areas.

The studies undertaken by Kamble, Aziz, Nelson, Rao, Misra, Vivekananda, Gupta, Rao and Ghosh spoke much about the comparison of the irrigation cost to total cost and also on per acre cost of cultivation of irrigated land. The studies have also analysed the impact of farm income, employment on production and distribution in the rural sector etc., etc.

A study of State Bank of India (1935), a study of the Agro Economic Research Centre (AERC) of the University of Delhi (1985) and a study made by Reddy (1988)[16] have focused their attention on the cropping intensity. The impact of minor irrigation schemes on other aspects like employment generation, rate of return on capital, cost aspects has not been revealed.

The study made by Aziz (1989)[17] has thoroughly analysed the impact of well irrigation on the area development and the cropping pattern. His thorough analysis has traced out the events that led to the spread of well irrigation and its impact on social and economic development of the village under study. However, his study has been confined to a particular village. Himisikote in Karnataka State, which restricted the scope of wider coverage.

A.N. Agrawall in his book *Indian Agriculture* (1980, Vikas Publishing House, Pvt. Ltd.), he feels that the cropping pattern in rain fed areas depends on the distribution of rainfall, temperature, climate, and nature of soil. Cropping pattern cannot be altered at will. The rapid expansion of irrigation facilities will enable the cultivators to put large areas under multiple cropping.

In 1986, an article entitled "The state of India's Environment 1984-85" in the *Second Citizens Report*, Centre for Science and Environment states that, a regional analysis of state-wise data reveals that tank irrigation is more prevalent in the southern part of the country than anywhere else. This southern region, consisting of the four states of Andhra pradesh, Tamil Nadu, Karnataka, and Kerala, account for 60 per cent of the irrigated area under tanks in the country. About 30 per cent of the net irrigated area in the southern region is under tanks.

The seminar on "Remote Sensing Applications in Renovation and Development of Irrigation Tanks" by P. Manavalan, P.P. Nageshwara Rao and S. Adige presented at Institute for Social and Economic Change, Bangalore, April 1998. This paper indicates a realistic appraisal of the use of satellite data as a survey tool for collecting useful information on various aspects of tank irrigation. The important results obtained from satellite based scientific study on revival of tank system in Kolar district, Karnataka are highlighted. It has been suggested to increase the adequacy and accuracy of the database on tank irrigation by suitably incorporating remote sensing techniques into field-based data collection programmes and to monitor by establishing the Natural Resources Information System for tank irrigation.

P. Thippaiah, in his paper "Status of Tank Irrigation in Karnataka" presented at workshop on Rehabilitation of Rain fed Irrigation Tanks with Farmers Involvement in Karnataka, held at ISEC, April 4, 1998. The study carried out Institute for Social and Economic Change has identified various causes/factors for the shrinkage of tank irrigated area by studying four tanks in Kolar district in 1996 where 32 per cent of the land was not cultivated. Although, the findings are relevant to only one sample district, these are generally applicable to the entire state as well as other states where tank irrigation is predominant. The important causes indicated by the study are:

- The storage levels of tanks reduced by 46 per cent due to siltation.
- The frequency of tank filling has also considerably declined.
- The physical and structural conditions of tanks are bad.
- Tanks are not irrigating the entire registered *atchkat* area due to excess use of water, improper management of water and absence of traditional *neeraganti* system.
- The sand excavation/mining in the foreshore area of the tanks is causing heavy infiltration of tank water, etc.

In order to make the tanks viable, some of these causes have to be deliberated and policy formulation should be based on the methods of adjustments. Priority should be given to rehabilitation and better management of tanks and tank water with the active participation of people. Otherwise, the water bodies on earth will slowly disappear causing for reaching effects.

In an article entitled "Efficiency, Equity and Environmental issues in Water Use" in *Margin* of July-Sept 2000. Prof. Devender B. Gupta is of the view that, in the absence of adequate and reliable water supply households are found to make their own arrangements for lack of access to adequate and reliable water supply. Our survey of the households have shown that households faced with these problems generally set up water storage facilities, dig wells, install tube wells, use lower quality of water from rivers, ponds, drains etc. boil water or set up water purifiers. For these they incur substantial costs. Crude estimates of the cost of such arrangements with provision of piped water supply services show that the cost differentials are pretty narrow. The first concerns are heavy initial investment costs needed to provide piped water supply. The other concerns the affordability aspect.

B.D. Dhawan, his article entitled "Drip Irrigation Evaluating Returns" in *Economic and Political Weekly*, Oct. 2000, states protective benefits of drip investment, as distinguished from its productive benefits, in the form of incremental output/income, have been fully recognized, particularly for farmers facing the prospect of declining water yield in their wells. Because of these twin benefits, such farmers have been more willing to try drip irrigation. It is time that these protective benefits are duly assessed in monetary terms and incorporated in to benefit cost estimates.

Dr. V.K.R.V. Rao in the concluding remarks of his book *Impact of Irrigation Studies of Canal, Well and Tank Irrigation in Karnataka* has narrated the impact of well irrigation, where he says, after the introduction of well irrigation, the per capita net income increased significantly. As a result, not only did the standard of living of the farmers go up above the average level of living that was reached in rural Karnataka, but also none of the farmers remained below the poverty line. This is a significant benefit which well irrigation has bestowed the farmers.

In an article in *Land Bank Journal* (March 2001) entitled "Impact of Minor Irrigation Loan on Farm Economy". H.O.Sharma, G.K. Pachoriya and T.B. Sing states, the introduction and increase in the minor irrigation loan and the facilities of irrigation created by these, increased the area under remunerative crops, reduced area under *Kharif* and *rabi fallow*, which increased gross cropped area and cropping intensity from 151.10 to 191.98 (40.88%). The introduction of minor irrigation loan also increased family, hired and bullock labour employment, and the net farm income and benefit cost ratio per hectare, to a considerable extent in all the cropping sequences and farms in the area under study.

M.V. Nadkarni in his book *Impact of Irrigation-Studies of Canal, Well and Tank Irrigation in Karnataka* (1979). According to M.V. Nadkarni and others study *Impact of Irrigation in Rural Areas*, irrigation explains 54 per cent of the total variance in agricultural production for India as a whole.

In an article entitled "Issues before Second Irrigation Commission of Maharashtra" in *Economic and Political Weekly* of March 24, 2001, R.S. Deshpande and A. Narayanamoorthy states Drip irrigation helps in avoiding the losses of water during conveyance and distribution. Available results in this regard show that water saving under drip irrigation ranges from 40 to 80 per cent compared to flood irrigation (INCTD, 1994). Apart from this, drip method brings in substantial improvement in the crop productivity. Among the various crops, all wide spaced and perennial crops like grapes, banana, coconut, mango, pomegranate, citrus, sapota and other horticultural crops as well as vegetables, have proved to be the most suitable crops for drip irrigation. However, it can be used efficiently even in the narrow spaced crops, though the life of the drip set reduces slightly under these conditions.

Studies have shown that drip irrigation even helped in reducing the cost of cultivation by reducing the losses in inputs like fertilizers, labours, tilling and weeding (inter cultivation) when compared with the convectional methods. Since fertilizers are supplied along with water (fustigation), the wastage of fertilizers through leaching and evaporation is much less. As a result efficiency of fertilizers increases significantly.

In an article in *Indian Farming* (August 2001) entitled "Decline of tank Irrigated Area in Kolar", Prof. P. Thippaiah of ISEC states, there is need to establish a National Tank Authority or Tank Board for proper identification of causes for the shrinkage of irrigated area, rehabilitation of tanks, management and utilization of tank water. More than this, the Centre and State governments have to take steps on war footing for the rehabilitation of tanks by involving village communities on the lines of PRADANs work in Madurai and Ramnad Districts of Tamil Nadu. These programmes should consist of desiltation, construction of silt arresting structures, contour bunding and afforestation in the tank catchments, eviction of encroachment of water spread area, supply channel and field channels, improvements to surplus weirs, slyices, field channels and regulation of bore wells.

In an article entitled "Dry land Agriculture and Rural Development" in *Yojana* of September 2001. Ishita Ganguli and Kamala Kanta Tripathy states, agriculture in India contributes nearly 30% of Net Domestic Product (NDP) and employs 70% of the people. Undoubtedly, agriculture is the backbone for the development of our economy. Out of 323 million hectares of the geographical area in India 123.6 million is cultivated. Out of this 123.6 million, only 40% of the land has fully or partially assured irrigation. The other 60% is either rain fed or non-cultivable.

In an article entitled "Expansion of Irrigation in West Bengal: Mid-1970s to Mid-1990s" in *Economic and Political*

Weekly of October 20, 2001, Vikas Rawal attempts to identify the basic structure of irrigation in West Bengal and the extent of growth of irrigation in the post-land reform period. His paper uses these data to identify certain features of irrigation development in the state. After presenting an overview of the pattern of irrigation in the state, the author analyses the growth of irrigation from groundwater and canals and the changes in irrigation from ponds and river lift-irrigation schemes.

S. Rajendran in his article "Drought in Karnataka" in *Economic and Political Weekly* of September 8, 2001, states, Karnataka is reeling under its worst drought in 10 years in the wake of the failure of the southwest monsoon. The symptoms were visible two months ago, but the state machinery is only now gearing up to tackle the situation, even as incidents of crop loss, cattle deaths, and mass migration of labour come to the fore. The state has sought financial assistance from the centre, but the question is how efficiently the money is spent and whether it reaches the needy. Ultimately, however, only long-term measures, such as rejuvenating the natural resource base, will help to tackle the scourge of drought.

In an article "Socio-Economic Constraints of Watershed Development Programme in Palakad district in Kerala" in *Agricultural Economics Research Review* of January-June 2003. T. T. Thomas, K. Jesy Thomas and E. K. Thomas states their study on the socio-economic constraints of the Watershed Development Programme in Palakad district showed that among the constraints faced by the beneficiaries, three constraints assumed greater importance, *viz,* non-availability of irrigation water, lack of technical guidance and lack of awareness of beneficial Programme. The analysis on strengths and weakness of the Programme indicated that co-operation of farmers was the major strength of the Programme. The problem of timely distribution of inputs was found to be an important weakness of the Programme.

G. Sastry, Y.V.R. Reddy, Om Prakash, C.A. Rama Rao and H.P. Sing their article "Evaluation of Watershed Development Programme in India" in *Agricultural Situation in India,* July 2003, states, Watershed Development Programme needs modification and introduction of new elements/procedures with stringent rules for punishment. Though, NGOs (well established) have been carrying this Programme, their staff is not well trained in this regard. However, many NGOs came into existence to avail the opportunity to utilise funds of this Programme have not been comparable with works carried out government officials who have to follow procedures or otherwise face consequences but this is not applicable to NGOs. Hence, funds collected by government need not be given to NGOs who can earn funds on their own capacity from donors/firms/private individuals, etc. to do social service. The works on water harvesting structures and vegetation development can be taken up on larger watershed basis considering the present status of such works.

In an article in *Indian Farming* (November 2003) entitled "Live Fencing System for Crop Protection and Natural Resource Conservation", M.S. Pendke and A.V. Gitte states, the farmers are reluctant to follow the agro-forestry as well as fencing techniques on their fields. The live fencing practices are adopted around orchards and irrigated crops but very rare in rain fed areas. However, extension workers had taken this job on large scale to popularize this live fencing system for soil and water conservation, crop diversification and sustaining productivity in rain fed region have been implemented over eight centers in Indian under National Agricultural Technology Project (NATP) to find suitable live fence species suited to particular region for rain fed farmers.

In an article entitled "Water Economy in Paddy" in Indian Farming of December 2003, B.P. Sing, R.P.S. Khokhar, D.D. Yadav and S.N. Verma states. In *Paddy*, the irrigation should be applied at 1.2 ratio (IW/CPE) or after 50mm (4-5 irrigations) or 8-10 days interval for gelting at par grain yield with irrigations at field capacity (10-11 irrigations). The 1.2 ratio not only yielded at par grain yield (5565 kg/ha) but gave maximum water use efficiency (7.48 kg/ha/mm of water) also.

Satyajit Singh in his book *Taming the Waters*, the political economy of large dams in India (*Indian Journal of Agricultural Economics*, October-December 2003) seeks to examine the role of irrigation, particularly through large dams, in contemporary Indian society. This study has also discussed the nature of Indian state as is reflected through its irrigation policies and discuss the challenge posed to it by emergent green movements. The book tries to focus much about the prices of environment, as it has control over water. Here the author makes a critique of large dams on the basis of their performance, ecological impact and unequal distribution of their costs and benefits in society. The author within a discipline of polities has carried out this work, but the methodology is characterized by its interdisciplinary nature. Thus the book focuses on ecological discussion, exploring the concept of sustainability with reference to irrigation, agriculture as well as people. The book has made an attempt to analyse irrigation in relation to two dominant ideas of equity and sustainability.

From the above reviews on the irrigation, it seems that they have covered the impact of different sources of irrigation like canals, tanks, wells, tube wells, river lift irrigation, pump sets and sprinklers. Further the studies have dealt with the different aspects of the impact of such scheme in terms of cost benefit ratio, financial rate of return, cost, cropping intensity, cropping pattern, employment, gross produce, resource use efficiency, etc. An analytical study about the impact of minor irrigation on the resource use efficiency, cropping pattern, employment opportunities, the household income of the farmers and on their standard of living in study area has not yet been attempted. Hence, this study is made to fill the gap in the literature on minor irrigation with special reference to tube well irrigation in study area.

Among the sources of minor irrigation like tanks, wells and tube wells, in recent years, tube wells have become more popular and the use of tube wells as a source of minor irrigation has acquired significance in most of the districts in Karnataka, Hassan district is one among them and this district is taken as the study area since tube wells are much popular in the interior rural areas where canal irrigation is not possible.

SCOPE OF THE STUDY

The need for irrigation in the agricultural development cannot be under estimated. Major irrigation works involve huge investment and take a long period for completion. But they have high capacity and can benefit large areas of land, sometimes running into millions of acres. Moreover, in recent years, major works are multipurpose projects are designed not only for irrigation but also flood prevention, navigation, hydropower, etc. The main advantages of minor irrigation works are that they require small investment, that they can be completed within a short period and their effect on agricultural production will be felt immediately. If quick returns are desired, minor irrigation works are very much suitable. Among minor irrigation sources tube well irrigation is definitely a superior method of irrigation system. However, for its successful application certain conditions are to be fulfilled. Therefore study concentrates only on one of the sources of minor irrigation, namely tube well irrigation.

THE ISSUES

The most important problem of farmers is selection of suitable tube well. Selection of tube well is dependent upon the level of sub-soil water table, structure of soil, the quantum of water required for irrigation, which, in turn, depends upon the area under cultivation, type of soil, type of crops grown, crop intensity, etc. and the rate at which water gets replenished in the tube wells.

Farmers are confronted with another problem, that is, installation of suitable tube well irrigation machinery. This machinery is liable to breakdown occasionally. Such breakdowns may affect the programme of watering plants very seriously, particularly when it would take a week or more to set things right. Crop field will dry up causing considerable damage to standing crops. Breakdowns arise not only on account of manufacturing defects and routine wear and tear, but also mostly on account of the ignorance of those who operate the machinery. Shortage of power supply and power cut severely affect the farmers, because on such occasions, power for irrigation pumpsets is supplied only on staggered time schedule. This causes great dislocation and hardship to farmers. Inadequacy of water yield is another problem of many farmers. Unless there is adequate water supply in the tube well its full advantage cannot be availed of.

Another most serious problem is the private exploitation of underground water by numerous individual farmers, which has tended to result in an indiscriminate and unregulated proliferation of tube wells, with consequential lowering of the water table in many tracts and increased cost of pumping. Further, affluent farmers, whose dependence on institutional credit has been much lower than that of smaller farmers, have escaped regulations based on distance criterion to which beneficiaries of institutional credit have been subject.

Development efforts such as digging of tube wells and installation of pumpsets require long-term credit repayable over a period of 5 to 20 years in easy annual installments. The agencies from which a farmer can borrow funds include moneylenders, government, commercial banks and cooperatives. Moneylenders constitute the traditional source of agricultural finance in Karnataka. They continue to play a dominant role in financing agriculturists. They charge exorbitant rates of interest and adopt various methods to extract the maximum from their debtors. Government also advances loans for purpose of agricultural development. The benefits of this assistance, however, do not always reach the small farmers, because they are unable to furnish the security as required by the rules and also cannot command any influence among the officials concerned. Commercial banks in the past, played a very insignificant role in providing finance to agriculturists. As their resources mainly consisted of short-term deposits and their lending policy was influenced by considerations of profitability and economy of operation, they did not prefer to finance agriculturists.

But, of late, the policy of socialization of commercial banks and inclusion of agriculture under priority sector for purpose of advances has prompted these banks to undertake financing agriculture in a big way. Co-operatives are also playing a significant role in financing agricultural development programmes. But they have certain problems such as inadequacy of resources and technical difficulties involved in verifying the title of the properties to be mortgaged by the borrowers leading to considerable delay in disbursement of loans.

Notwithstanding the above stated problems prevailing in the study area, the present study is intended to evaluate the impact of tube well irrigation on agricultural production with special reference to paddy cultivation in Hassan district of Karnataka.

OBJECTIVES

The study aims at the following objectives.

1. To study the growth and pattern of tube well irrigation in study area.
2. To analyse the cost of production of paddy with reference to tube well irrigated farmers.
3. To analyse the performance of the productivity of paddy cultivation with reference to the tube well irrigated farmers.

STUDY AREA

The present study covers the Hassan district in Karnataka for investigation on the basis of the following reasons.

1. Tube well irrigation dominates in the district.

2. It is an important district where ground water is being exploited at faster rate.
3. Among minor irrigation sources, tube well irrigation have become popular and they are given more attention since other sources like canals and tanks depend mostly on rainfall.

In study area, Hemavathi is the only perennial river. Since there is frequent failure of monsoon, the availability of water from rivers and reservoirs is not sufficient. Moreover, all tanks are rain fed. So the only source of irrigation, which provides perennial water supply for cultivation, is tube wells. The present study has been undertaken to investigate the functioning of tube well irrigation in agriculture with special reference to paddy cultivation.

REFERENCE PERIOD

The reference period of the present study is 2006-2007. The size of land holdings, cost of cultivation, gross and net value of output, marketed surplus, etc. are indicated with reference to this point of time. Where, however, a attempt has been made to trace the changes in the cropping pattern, output, etc. the reference period is 2005-2006.

CHAPTER SCHEME

The work is presented in 5 chapters. Chapter 1 is the Introduction, which consists of the importance of irrigation, role of irrigation, impact on agriculture, weakness of irrigation, need for minor irrigation, review of literature, scope of the study, objectives, study area and reference period, chapter 2 contains, data source, a profile of study area, sample design, method of investigation, date of field enquiry, statistical techniques, limitations, concepts and definitions.

The chapter 3 gives a picture of the development of irrigation in the India as well as in the Karnataka. Chapter 4 reveals the impact of tube well irrigation on agricultural production with special reference to paddy cultivation in study area.

A summary of major findings of the study, suggestions for accelerating agricultural development through tube well irrigation and factors to be reckoned with in further research studies in this area are given in the last chapter.

Thus the study investigates and attempts to throw light on the problems and prospects of existing sources of tube well irrigation with special reference to paddy cultivation in Hassan district of Karnataka.

REFERENCES

1. *Environmental Encyclopedia on Irrigation*, 2007.
2. *Indian Economy*, A.N. Agarwal.
3. The Five Year Plan for Bombay (1951-52–55-56), Directorate of Publicity Govt., of Bombay, p. 3.
4. "The irrigation colonies have in fact opened an era of prosperity undreamt of in the past", M.L. Darling, Punjab peasantry in prosperity and Debt. 1932, p. 119.
5. Knowles, *Economic Development of British Empire Overseas*, Vol. I, pp. 367-68.
6. V.S. Vyas, "Some Aspect of Irrigation and Power Development, in Vadilal Daglli (ed). *Foundations of Indian Agriculture* (Bombay 1948), p. 139.
7. B.D. Dhawan, "Indian Irrigation: An Assessment," *Economic and Political Weekly*, May 7, 1988, p. 965.

8. B.D. Dhawan, "Irrigation in India's Agricultural Development" (unpublished paper 1998). Quoted in Ashok Gulati and Sudha Narayana. *The subsidy syndrome in Indian Agriculture* (New Delhi, 2003), p. 157.
9. As noted by the Report on Currency and Finance, Reserve Bank of India, 2000-01, para 3.37.
10. C.H. Hanumantha Rao, Susanta, K. Ray and K. Subba Rao, *Unstable Agriculture and Droughts* (New Delhi, 1988), pp. 49-55.
11. B.D. Dhawan (1988), *Op. cit.* p. 966.
12. Report of the Indian Famine Commission, 1898, Government of India, Agricole Publishing Academy, New Delhi, 1979, p. 331.
13. Evaluation of Ground Water Irrigation in Kota District. Rajasthan (An Ex-Post) ARDC, Programming and Evaluation Division, Bombay, 1982. pp. 7-40.
14. Evaluation Study Series No. 12, "Minor Irrigation in Bhojpur District, Bihar". (An Ex-Post Evaluation Study) ARDC, *op. cit.*, pp. 28-39.
15. Evaluation Study Series No. 19, "*Public Tube Wells and River Lifts in Orissa-An Ex-Post Evaluation Study*", Economic Analysis and Publication Department, NABARD, Bombay, 1984, pp. 1-9.
16. An Ex-Post Evaluation Study of Direct Irrigation (DEEP) Tube Wells in Blaspur Block of Ambala District of Haryana-*National Bank News Review*, November 1985, p. 31.
17. Abdul Aziz, *op. cit.*, pp. 2-15.

2

RESEARCH METHODOLOGY

The present study concentrates upon a study of the impact of tube well irrigation on agricultural production with special reference to paddy cultivation at two villages Dammaningala and A. Kalenahalli from Channarayapatna taluk of Hassan district. This study was carried out by an empirical investigation conducted by canvassing a structured questionnaire, which was supplemented by an unstructured questionnaire. However, secondary data, wherever available, was made use of. The various aspects concerning Methodology have been proposed to describe under the following heads.

DATA SOURCE

The study is based on both the primary and secondary sources of data. The primary sources are the villagers, comprising of tube well irrigated farmers. A pilot study has been conducted before actually commencing the primary survey. The field investigation and data collection were carried out during the period from April 2007 to July 2007. The primary data pertaining to paddy cultivation, land use, cost of production, use of inputs, income, asset, employment, etc. were collected directly from the tube well owners of the sample villages.

The secondary sources of information and data include the documentary sources such as the Census, Irrigation Information Reports, Reports of the Bureau of Economics and Statistics, Agricultural Statistics and Village Records, Assistant Director of Statistics Department in Hassan, Govt. of India publications on Irrigation, Economic Survey-Karnataka, News papers, Journals and books.

A PROFILE OF STUDY AREA

Hassan, which is one among the 28 districts of the Karnataka State, is situated in the south-western part of the State. It constitutes 8 towns, 8 *taluks,* 38 *hobli* hqs, 359 *Gram panchayat,* and 2393 villages with populated and 163 non-populated villages. The district has had an eventful and rich history. In the past, it reached the height of its glory during the rule of the Hoysalas who had their capital at Dorasamudra, the modern Halebid in Belur *taluk.* The district, noted for its enchanting natural scenery of *malnad* (a mainly forested, hilly region of heavy, reliable rainfall) is also a veritable treasure house of the Hoysala architecture and sculpture, the best specimens of which are at Belur and Halebid. Shravanabelgola, in Channarayapatna *taluk,* which is studded with Jaina monuments, is a internationally renowned centre of pilgrimage for the Jains.

The district, situated between 12^0–31N and 13^0–33N–north latitude and 75^0–38N and 76^0 –39N–east longitude, is in the south-western part of the State. The greatest length of the district, from north to south, is about 80 miles or 129 kilometres, and its greatest breadth, from east to west, is about 72 miles or 116 kilometres. It is bounded on the north by Chikmagalur district, on the east by Tumkur ad Mandya districts, on the south by Mysore and Coorg districts and on the west by South Kanara district.

The geographical area of the district, according to the Central Statistical Organization of the Government of India, is 6,845 square kilometres, which works out to 2,645 square miles. But the reporting area of the district for land utilisation purposes, as worked out by the State Survey, Settlement, and Land Records Department, is 6,940.4 square kilometres or 2,679.7 square miles. This slight difference is due to the different methods employed by them in measuring the area.

The population of the district, according to the census of 2001, was 17,21,669 of which 8,59,086 were males and 8,62,583 were females. Hassan is one of the smaller districts in the state. In respect of area, as per the figures of the Central Statistical Organization, it occupies the 15th place among the districts of the State. The density of population in the district in 2001 was 251 per square kilometre and this was below the State average, which were 275 per square kilometre.

The average annual rainfall in the district is 1031 millimeters. Most of the rainfall in the district is confined to the period from May to October, July being the rainiest month.

The economy of the district is primarily based on agriculture, which is the chief occupation of the people as in other contiguous districts. As the major agricultural operations have to synchronise with the rains, the agricultural year is divided into four seasons, *viz.*, south-west monsoon period from June to September, north-east monsoon period or retreating monsoon from October to December, cold–weather period from January to February and summer season or hot-weather period from March to May. The hot-weather rains which come in the form of short-lived tempests are very important to successful farming; they feed a number of tanks. These rains also enable the cultivators to prepare the lands for the following south-west monsoon season in the *maidan* parts and for setting up of coffee plants in the *malnad* parts. It is during the south-west monsoon period that the ploughed fields receive the seeds. During the cold weather period, the fields are reaped and the grains stacked. The north-east monsoon which comes in heavy showers is of great value as it brings plenty of water to the tanks in the *maidan* parts. The wet cultivation from the month of December upto March and April depends entirely on the north-east monsoons.

There are two main harvest-the *kharif* known as *mungaru* and the *rabi* known as *hingaru*. Both the important food-crops, *ragi* and paddy, are *mungaru* crops for which ploughing starts in April, while sowing is done with the first rains of the south-west monsoon, usually by the end of May and the beginning of June and sometimes extending up to July: transplantation is done from July to September and harvesting and threshing from November to February. Oilseeds, cotton and most of the pulses are also grown as *mungaru* crops.

The district had 460 thousand hectares of land under cultivation in 2004-05, which constituted about 69 per cent of the geographical area of the district. In 2004-05, the net area irrigated in the district from all sources was 92,976 hectares. Out of this, an area of 31,974 hectares was irrigated by canals, 22,133 hectares by tanks, 1,188 hectares by wells, 36,600 hectares by tube wells, 19 hectares by lift irrigation and 1,062 hectares by other sources. On the basis of 2004-2005 statistics the total number of tube wells are 1261 in Alur taluk, 1541 in Arkalagud, 2418 in Arsikere, 1788 in Belur, 2124 in Channarayapatna, 1701 in Hassan, 1292 in Holenarasipur and in 924 in Sakaleshpur (Manjarabad) and there are totally 13022 tube wells are in Hassan district.

The three main rivers of the district, the Cauvery, the Hemavathy and the Yagachi, forms the main sources of surface water for the several canals of the district drawn from the anicuts built across them. Canal irrigation is mostly in the *taluks* of Holenarsipur, Arkalgud and Channarayapatna. Tube wells provide access to the subterranean water or to the waters of the rivers, which run deep below the level of the ground. Importance of tube well irrigation, especially in the *maidan* parts of Arkalgud, Arsikere, Holenarsipur and Channarayapatna *taluks* of the district is considerable.

Ragi and *Paddy* are the most important cereals of the district and they are grown in all the *taluks*, while the acreage under ragi in Manjarabad taluk is negligible; the acreage under rice in Arsikere taluk is megre. Jowar is grown mainly in Arsikere taluk. While Arsikere and Belur are noted for coconut cultivation, sugarcane is grown mainly in Belur, Hassan and Channarayapatna taluks. While cardamom is almost confined to Manjarabad taluk, coffee is grown in Manjarabad, Hassan and Alur taluks.

In Hassan district, Hemavathi is the only perennial river. Since there is frequent failure of monsoon, the availability of water from rivers and reservoirs is not sufficient. Moreover, all tanks are rain fed. So the only source of irrigation, which provides perennial water supply for cultivation, is tube wells. The present study has been undertaken to investigate the functioning of Tube well irrigation and find answers to the following questions in the district.

1. Whether is there any change in the pattern of tube well irrigation in the district?
2. Whether it has generated employment and income to the farmers?
3. Whether Tube well irrigation has any impact on the socio-economic conditions of Tube well owners? and,
4. Whether the standard of living of tube well owners has improved?

Hence, it is felt that a detailed study is essential to know the various intricacies faced by the tube well irrigation and suggest suitable solutions to solve them. And there is an ample scope for research on the above topic.

SAMPLE DESIGN

The present study is carried out in Hassan district of Karnataka. A three stage random sampling technique was adopted to select the *taluk*, villages and the farmers. Accordingly, out of the total of 8 *taluks* in the district, one *taluk* (Channarayapatna) was selected as first stage unit of sampling representing tube well irrigated taluk on the basis of selected economic indicators.

For the selection of sample villages, the second stage unit, all the villages are listed out and two villages (Dammaningala and A. Kalenahalli) under selected *taluk* were selected using random sampling procedures. All the farm households in the selected villages are listed out.

From each sample village 50 households are selected in probability proportion to the existing number in each category in the selected village. Thus, the study covers 100 sample households, which were selected for the primary survey considering their experience in irrigated agriculture, land holding sizes and other relevant characteristics.

METHOD OF INVESTIGATION

In all, this gave us a total of 100 respondents comprising 50 from tube well irrigated farmers from Dammaningala village and 50 from tube well irrigated farmers from A. Kalenahalli village of Channarayapatna taluk in Hassan district. The respondents were approached through the village level worker and village accountant. After establishing rapport with them, the investigator filled in the questionnaire.

STATISTICAL TECHNIQUES

The study uses a number of statistical techniques in presenting the analytical results. The most important of the techniques used are the averages, percentages, and ratios, mean variations and benefit cost analysis, etc. Graphs, diagrams and flow charts are taken recourse in presenting facts and figures on the study area.

DATA ANALYSIS

After scrutiny, the collected data were processed to fulfill the objectives laid down for the study. The processing of data was done with help of a computer, a calculator and also manually. Simple statistical techniques such as percentages, multiple regression, etc. were used.

The significance of the results obtained from the various statistical and econometric techniques were statistically tested.

LIMITATIONS

The lack of maintenance of records relating to input and output by the farmers and also the deliberate concealing of facts and figures by the farmers are the limitations, particularly when survey method of investigations is used. Though every effort will be made to elicit correct information by careful probing, it is obvious that lapse of memory on the part of the respondents regarding details about operations and resource utilisation could not be completely overcome. However, much care will be taken to collect reliable data by probing further.

CONCEPTS AND DEFINITIONS

This study used standard concepts and definitions. Brief account of these are given below.

Tube Well Irrigated Farmers: A farmer who owns irrigated land under tube well irrigation and whose major source of income is agricultural production.

Family Labour: It consists of actual work carried out by the family members for crop production. This labour has been values on the basis of prevailing wages paid to the hired labour.

Hired Labour: Hired labour includes permanent as well as casual labour. The evaluation of both has been done on the basis of the prevailing wage rate in the village.

Bullock Labour: This item includes both hired and own bullock labour. Owned bullock labour has been accounted as per the rates of hired bullock labour in the locality.

Seeds: Purchased seeds have been valued at actual cost and farm produced seeds have been evaluated at prevailing village prices.

Manures: Produced manures have been evaluated at prevailing village market prices. In the case of purchased manures, fertilizers and pesticides, the actual cost has been taken into account.

Land Rent: Rental value of owned land has been considered on the basis of local practice. In the case of land taken on lease, the land rent actually paid is accounted for.

Depreciation: The depreciation was calculated by straight-line method at the rate of 2 per cent in the case of permanent buildings and tube wells and 5 per cent in case of electricity motors and other temporary buildings.

In case of implements, the depreciation was calculated at the rate of 15 per cent on ploughs and other implements while the depreciation on bullock carts was 5 per cent, and 5 per cent on tractor and attachments. Minor repairs were directly added to the depreciation while major repairs were treated as addition to the value of implements and depreciation on its value was worked out in the usual way.

Cost of Cultivation: The analysis of cost of cultivation of crops was based on the Direct Costs and Indirect Costs. These two costs include the following items.

(a) **Direct Costs:** It includes the cost items such as;

I. Hired labour

II. Bullock labour/tractor

III. Seed

IV. Manure

V. Fertilizers

VI. Pesticides

VII. Irrigation charges

VIII. Miscellaneous

IX. Land revenue

(b) Indirect Costs:

I. Interest on working capital

II. Depreciation

III. Interest on fixed capital

IV. Rental value

V. Family labour

Measures of Income: Farm business analysis has been carried out by using different measures of income as shown below:

(a) **Gross Income:** Value of main produce and by-produce were calculated at the prevailing post-harvest prices in the area profits at different cost concepts used above, provide different estimates of returns to the sample cultivators and have been worked out on per acre basis as detailed below.

(b) **Farm Business Income:** Farm business income is the difference between the gross income and direct costs (i.e., profit at direct costs).

(c) **Net Income:** The difference between gross income and total costs represents the net income from the farm enterprise.

Cost of Production for Joint Product: The paddy crop consists of two saleable commodities viz., grain and its by-product. Since the cost of production is available for the crop as a whole, the cost has been apportioned between the main produce and the by-produce in proportion of their share in the gross income. It facilitates working out the cost of production per quintal of the main produce.

The detailed methodology of the survey is presented in Chapter 4, which mainly deals with the impact of tube well irrigation on agricultural production with special reference to paddy cultivation in study area.

3

IRRIGATION DEVELOPMENT IN INDIA

Irrigation in India has been practiced from ancient times; and to supplement and conserve the rainfall.[1] Provision of water for cultivation had figured prominently among the duties enjoined on irrigation tanks and wells are a familiar feature of the Indian landscape the rulers of the land, who undertook the construction of irrigation works as benevolent works and many of the works of the ancient time owe their existence to the munificence of kings and philanthropists.

ANCIENT PERIOD

Ancient Indian civilization mostly developed in the river valleys, which were well equipped with irrigation system, which helped in the growth of food crops and cotton. *Vedas* refer to *avatar* or water wells, *Kulya* or canal, and *Sarsi* or dam indicating the fact that devices for irrigating land were already known. Manu mentions about *tataka* or artificial storage. *Mahabharata* (Epic) contains indications of the practice of irrigation from wells, tanks and canals. Kautilya observed, "If privately managed dams are neglected for 5 years, their charge is taken over by the State. If they are constructed by public contribution, revenue is to be remitted for 5 years. If only repairs are carried out by public effort revenue is to be remitted for 4 years." Classical literature is replete with water courses-*pranadi, kulya, nala, naliha, tilamaha,* etc.

Artificial lakes and canals that dot the country in hundreds are centuries old, and some o them have served for more than a thousand years. Megasthenes (4 centuries B.C) mentions that the whole country is under irrigation; and very prosperous because of the double harvests, which they were able to reap each year because of irrigation. "Storages and their water courses are common in south, central and eastern India. The first structure, a weir built in stone and clay, was probably laid in the 2nd century A.D. This Grand Anicut was built across the river Kauvery, about 330 m. long, 12 to 18 m. wide and 4.6 to 5.5 m. high. The ancient has serviced irrigation withstood tne annual Kauvery floods for more than 1600 houıs".

MUGHAL PERIOD

Irrigation development received great attention during the pre-Mughal and the Mughal periods. The Western Jamuna Canal (of Firoze Shah Tuglak) brought water to Delhi in the 14th century. It served as an irrigation canal in the tract it traversed. The Bari Boab canal was executed by Ali Mardan Khan in the middle of the 17th century. The Eastern Jamuna Canal was laid down in the reign of Mohammed Shah during the middle of 18th century.

BRITISH PERIOD

The British inherited a tradition of irrigation in India. The Indian Irrigation Commission of 1901-03 recorded,

"Be this as it may, it is certain that it was the existence of the grand Anicut in Madras, and the remains of old Mohammedan channels in the Punjab and United Provinces, which suggested and led to the construction of the earliest works carried on under British rule. India, therefore, in a great measure owes to her former rulers the first inception of the present unrivalled system of irrigation works."

Initially, during the British period, Western Jumuna and Eastern Jamuna canals were renovated and remodeled followed by the reconstruction of the Kauvery Delta system in conjunction with the Grand Anicut. A masonry weir across the coleroon was constructed in 1836. The upper Ganga canal was commenced in 1842 and completed in 1854 by P. Cautley. This was followed by other canals like the Lower Ganga, the Agra and the Betwa canals in U.P; the Sirhind canal in the Punjab, Mutha canal in Bombay; and the Periyer canal in South India. Other notable works were the weirs across the river Godawari near Rajah mundry, and across the river Krishna near Vijayawada; along with their canal systems, which led to the irrigation of the fertile Godawari- Krishna deltas.

Occurrence of very serious famines towards the later part of the 19th century led to the development of the Indian Irrigation Commission in 1901, to report on irrigation as protection against famines in India. This gave further impetus to irrigation projects, and led to a number of projects like the Triple Canal Project in the Punjab; the Godawari canal in Bombay and the Tribeni canal in Bihar.

Further irrigation works were taken up after the First World War, notable example being the Mettur Nizamsagar and Krishna Raja Sagar projects; the Periyer canal the Khadakvasla storage dam. Pravara and Nira canal in the

South and the Ganga canal, the Sarda canal in the North. Besides these major systems, a large number of medium and small irrigation works like tanks and canals were also constructed all over the country. This led to substantial increase in irrigation and food production.

POST INDEPENDENCE PERIOD

During the depression of 1930s and the Second World War, little development took place in the field of irrigation. With the attainment of Independence in 1947 the demand for food production led to the launching of many irrigation projects. The partition of the country resulted in major irrigated areas going to Pakistan. Of the total volume of water carried by the canals in undivided India the canals in Pakistan received 81,400 m.cu.m. Of water against 11,000 m. cu.m. of water used by India in the Indus basin. In terms of irrigated area about 8 m. hectares of land went to the share of Pakistan, against only about 2 m. hectares left in India. This further necessitated the growth of irrigation potential.

PROGRESS OF IRRIGATION IN INDIA

Soon after Independence, the Government gave great importance to the development of irrigation works to encourage the farmers to produce more. The recent developments in respect of major works have been in the direction of building dams across the rivers and storing water in large reservoirs, from which canals are taken out for irrigation. Most of these projects are known as multi-purpose works; for, they are not only meant to provide irrigation, but also aimed at controlling the flow of the rivers so as to avoid floods and to generate electricity by using waterpower. Since 1951, a number of multi-purpose projects have been completed and many others are in the process of construction. The most famous multi-purpose projects are Domodar Valley project to benefit West Bengal and Bihar, Bhakra Nangal project to benefit Punjab, Haryana and Rajasthan, the Hirakud Dam project of Orissa, the Chambal project, the Rihand Project, the Kosi project in North Bihar, etc. The progress of irrigation since 1950-51 is given in the following table.

TABLE 3.1

GROSS AND NET IRRIGATED AREA IN INDIA

(Million hectares)

Year	*Net Irrigated Area*	*Gross Irrigated Area*	*Total Cropped Area*	*Gross Irrigated area as per cent of sown area*
1950-51	21	23	133	17
1970-71	31	38	166	23
1990-91	48	62	186	34
2000-01	57	76	193	39

Source: Agricultural Statistics at a Glance (2002)

Above table gives us an idea of the progress of irrigation since 1950-51 when economic planning was undertaken:

(a) Net irrigated area has increased from 21 million hectares in 1950-51 to 57 million hectares in 2000-01- an increase by over 170 per cent in 5 decades.

(b) Gross irrigated area (which includes land cultivated more than once in a year with the help of irrigation) has gone up from 23 million hectares to 76 million hectares during this period—an increase by more than 230 per cent in 5 decades.

(c) This means that the area cultivated more than once in a year has risen sharply from about 2 million hectares in 1950-51 (which was 8 per cent of net irrigated area) to nearly 19 million hectares in 2000-01 (which was 33 per cent of net irrigated area). This has great significance in the Indian context. Area irrigated more than once is a kind of increase of land and, therefore, is very crucial in raising agricultural productivity and agricultural production.

(d) Finally, the benefit of irrigation is available for more land now than in 1950-51. For instance, the third column in Table 3.1 shows that in 1950-51, total cropped area was 133 million hectares; but by the end of the century, the total cropped area had increased to 193 million hectares. Moreover, only 17 per cent of cropped land got the benefit of irrigation in 1950-51. But now as much as 39 per cent of all cropped area gets the benefit of irrigation.

When India started planning in 1950-51, the irrigation potential from major and medium irrigation was about 10 million hectares and from minor irrigation works was 13 million hectares, totaling 23 million hectares. By 2001-2002, the total irrigation potential created had increased to 95.40 million hectares.[2] Ministry of Water Resources has recently estimated that India's ultimate irrigation potential from major and medium works increased to 59 million hectares and minor works contributed 81 million hectares making a total of 140 million hectares. By trans-basin diversions it is estimated that another 25 million hectares potential can become available through surface and 10 million hectares through growing water sources.

The present long-term objective is to utilize the available water resources fully by the year 2010. For this purpose, the Government has prepared a national perspective for water resources development which includes also storage and inter-basin transfer from surplus to deficit and drought-prone areas.

The irrigation potential through major, medium and minor irrigation projects has increased from 22.6 million hectares (mha) in 1951, to about 102.77 mha at the end of 2006-2007. The development of irrigation potential since 1950-51 is given in the following table.

TABLE 3.2

DEVELOPMENT OF IRRIGATION POTENTIAL

(Million hectares)

Year	*Major and Medium works*	*Minor works*	*Total*
1950-51	10	13	23
1980-81	27	31	58
1990-91	30	48	78
1999-00	35.35	59.38	94.73
2005-06	38.60	59.50	98.10
Ultimate potential	73.0	66.8	139.89

Source: Economic Survey, 2006-07

The table shows clearly how the land under irrigation is increasing steadily. Between 1950-51 and 2001-2002 irrigated area increased from 23 million hectares to 96 million hectares or by 322 per cent. The utilization has also increased by more than three and a half times (from 22.6 million hectares to 84.70 million hectares) over the same period. However, as against the ultimate irrigation potential of 139.89 million hectares, the actual utilization at the end of 2004-2005 was only 83.41 million hectares, which is only 60 per cent of the potential. Plan wise irrigation potential created and utilized through major, medium and minor irrigation projects in the country is as indicated hereunder.

TABLE 3.3

PLAN-WISE POSITION OF IRRIGATION POTENTIAL

Plan Period	*Potential (Cumulative) created (Mha)*	*Potential (Cumulative) Utilized (Mha)*
Pre-Plan period	22.60	22.60
First Plan (1951-56)	26.26	25.04
Second Plan (1956-61)	29.08	27.80
Third Plan (1961-66)	33.57	32.17
Annual Plans (1966-69)	37.10	35.75
Fourth Plan (1969-74)	44.20	42.19
Fifth Plan (1974-78)	52.02	48.46
Annual Plans (1978-80)	56.61	52.64
Sixth Plan (1980-85)	65.22	58.82
Seventh Plan (1985-90)	76.53	68.59
Annual Plans (1990-92)	81.09	72.86
Eighth Plan (1992-97)	86.26	72.24
Ninth Plan (1997-2002)	93.95	80.06
Tenth Plan (2002-2007)	102.77	86.29

Source: India 2008, A Reference Annual, Published by Ministry of Information and Broadcasting, Govt. of India.

This shows that by proper planning and concerted efforts it is possible to increase the area under cultivation substantially. This alone can considerably increase agricultural production and productivity.

SOURCES OF IRRIGATION IN INDIA

Sources of irrigation in India can be divided into: (i) Wells, (ii) Tanks, (iii) Tube Wells, and (iv) Canals.

1. Wells

The wells are the oldest and the most important source of irrigation in India. These wells may be either *kuchcha* or *pucca*. There is a large number of *kuchcha* wells in our country as they are very cheap to construct and do not need any technical skill. Water from these wells is raised through *dekhli*. The *pucca* wells are of two types-ordinary wells, from which water is drawn through the Persian wheel or *charas* and tube-wells from which water is raised by means of electric or diesel pumps.

Wells are usually constructed in planes and other areas where the water-level is not very deep and the soil is soft. They are used for irrigation purposes in Punjab, U.P., Bihar, West Bengal, the eastern side of Western Ghats and the black-cotton soil area. They are both private and Government owned.

Irrigation through wells is useful as the water is rich in soda nitrates, chlorides and sulphates which make the land more fertile. Moreover, there is no danger of water-logging as in the case of canal irrigation. It would be better if individual farmers are encouraged to construct more *pucca* wells by the grant of *takavi* loans. Boring facilities should be made available to them and they should be supplied with electric power for installing power pumps, which would be cheaper and better. The construction of new wells for purposes of irrigation can be taken up by co-operative societies to be formed for that purpose.

2. Tanks

Irrigation through tanks is more common in the south, particularly in Tamil Nadu and Karnataka, though there are some tanks in Bengal, Bihar, Orissa, U.P. and Madhya Pradesh. Tamil Nadu has the largest number of tanks. These tanks vary in size from big lakes to small village pools and there are both artificial and natural tanks. Artificial tanks may be either *kuchcha* or *pucca*; the *pucca* tanks are constructed both below the surface and above water during the non-rainy season.

Like well water, tank-water is also very fertile as it contains rain water and filth. Tanks are an important source of irrigation but usually their condition is very unsatisfactory. The tanks gradually get filled up with earth. This should be checked by deepening the old tanks. New tanks should be built to take the place of old ones, which have gone out of use. Since the individual cultivator cannot do much in this direction, the State should take a greater interest in the digging of tanks, as is being done by the U.P. Government since 1948. This work should be entrusted to *panchayats* and other local bodies and to co-operative societies. The construction of *pucca* tanks, though a little costlier, should be encouraged as the water stored in them can be retained for a longer period. Water supply from these tanks can be improved by constructing *pucca* pipe lines from the tanks to the fields thereby preventing waste. Moreover, when these tanks become empty, canal water can be diverted to them during periods when it is not needed for irrigation purposes.

3. Tube Wells

These are sunk by inserting pipes below ground surface and passing through different geological formations of water bearing and non-water bearing strata. Blind pipes are located at non-water bearing strata and perforated pipes at water bearing strata.

Tube wells are common in areas where the water table is rather deep, say over 15 metres. The sub-soil water is exploited through deep well pumping.

Tube wells have been successfully constructed primarily in the upper reaches of the Indo-Gangetic valley and in certain coastal deltaic areas, which satisfy the following conditions:

- The flow of water in the subsoil is adequate to meet the surface demand thus ensuring a stable water table;
- The depth of the water table below the ground does not ordinarily exceed 50 ft;
- The demand for irrigation prevails over a wide area for an average period of not less than 3,000 hours in the year; and
- For lifting the water electricity is available over the tract which economises and popularizes lifting operations.

The tube wells have an advantage over the canals in that the water can be taken directly at the places required and the flow on the spot takes place by gravity. In offering results, these are superior in performance and of value to a surface well. A tube well irrigates about 100 acres and the cost of a tube well, with complete accessories is between Rs. 50 to 80 thousands depending upon whether electrical energy is available for running the pump or not. This is for a 300 ft. deep tube-well, discharging 30,000 gallons per year.

Further, a tube-well is particularly suited to areas of small holdings where construction of wells raises difficult problems of finance. Also, in periods of draught, the tube well continues to give an unabated supply of water. Therefore, it is possible to do crop planning on dependable and successful pattern and to carry out a programme of promoting scientific agriculture on such units as tube wells provide.

4. Canals

Canal water is a major source of irrigation in India and has been in vogue from ancient times. Canals are to be found in Punjab, U.P., Bengal, Bihar, Tamil Nadu, Karnataka, Andhra Pradesh, Maharastra, Madhya Pradesh and Orisa. There are two types of canals–the perennial canals and the inundation canals. The perennial canals are constructed by storing water in artificial dams which is available throughout the year. In the inundation canals water accumulates when the rivers overflow their banks. The former type of canals are better as they also serve in periods of drought when the latter type of canals become unworkable, their water having dried up. The cost of construction of inundation canals is much lower than the cost of perennial canals, which need technical skill, machinery and material for their construction.

There are certain disadvantages of canal irrigation. It gives rise to the problem of water-logging, particularly in Maharastra and Punjab. Canal irrigation results in the loss of a large quantity of water before it finally reaches the fields. These difficulties can, however, be overcome by properly lining the canals.

Approximately 32 per cent of the irrigated areas in India are watered by canals. Canal-irrigated area has increased from 8.3 million hectares to 18.0 million hectares during 1950-51 and 1999-200. Even then, its relative importance has come down from 40 per cent to 31.5 per cent.

During this period, it is well irrigation, particularly tube well irrigation which has made the most spectacular progress. In 1950-51, only 0.1 million hectares were irrigated by tube wells, but in 1999-2000. About 16 million hectares were served by tube well irrigation. Tube wells account for 28 per cent of total irrigated area. Well-irrigated area has increased from 6 million hectares to 34 million hectares during the last 50 years–well irrigation in 2000-2001 accounted for 59 per cent of the total irrigated area as compared to only 29 per cent in 1950-51. The growth of well irrigation has been at the expense of irrigation from tanks and other sources. In 2000-2001, tanks watered 4.7 per cent or a total irrigated area.[3] The area irrigated by sources in India since 1950-51 is given in the Table 3.4

TABLE 3.4

AREA IRRIGATED BY SOURCES IN INDIA

	1950-1951		2000-2001	
Sources of Irrigation	*Area (Million hectares)*	*Percentage*	*Area (Million hectares)*	*Percentage*
Canals	8.3	40	18.0	31.5
Wells and Tube wells	6.0	29	33.6	58.7
Tanks	3.6	17	2.7	4.7
Other sources	3.0	14	2.9	5.1
Total	**20.9**	**100**	**57.2**	**100**

Sources: Central Statistical Organization (CSO) Statistical Abstract 2002.

IRRIGATION AND GOVERNMENT POLICY

At the initiation of planning in India in 1950-51, irrigation schemes were divided in to three categories- major, costing more than Rs. 5 crore each; medium, costing individually between Rs. 10 lakh and Rs. 5 crore each; and minor, costing less than Rs. 10 lakh each. A new classification was adopted in April 1978. According to this classification, major schemes are defined as those having culturable command area (CCA) of more than 10,000 hectares, medium schemes as those having CCA between 2000 hectares and 10,000 hectares and minor schemes as those having CCA of less than 2,000 hectares.

Massive investment was made on irrigation during the planning period. From Rs. 455 crore in the First Plan, the expenditure on irrigation rose to Rs. 36,649 crore in the Eighth Plan. The outlay for irrigation and flood control in Ninth Plan was Rs. 63,682 crore. Of this Rs. 48,259 crore (76%) was for major and medium irrigation projects and the balance Rs. 15,423 crore (24%) for minor irrigation, Command Area Development (CAD) programme and flood control. The outlay on irrigation and flood control in the Tenth Plan (2002-07) has been kept at Rs 1,03,315 crore.

Substantial expenditure has gone in for developing the major and medium irrigation potential especially the major river valley projects like the Bhakra Nangal Project (Punjab), Beas Project (Punjab and Haryana), Hirakud Dam project (Orissa), Damodar Valley Corporation (Jharkhand and West Bengal), Nagarjunasagar Project (Andhra Pradesh and Karnataka), etc. However, minor irrigation continues to occupy an important place as its share in the total irrigation potential of 93.95 million hectares in 2001-2002 was 61 per cent (56.90 million hectares). This shows that minor irrigation accounts for a little less than two-thirds of irrigation potential in the country.

Introduction of the new agricultural strategy and high-yielding varieties programme required substantial irrigation facilities. Consequently, the Fourth Plan recommended early completion of old schemes and introduction of new schemes. The Fifth Plan introduces a comprehensive programme and 38 command area development authorities were set up covering 50 irrigation projects. Total Irrigation potential created till the end of the Ninth Plan (1997-2002) was 95.40 million hectares[4] as far as irrigation programmes are concerned. Rural Infrastructure Development Fund (RIDF) was launched in 1995-96 with a corpus of Rs. 2, 000 crore to provide loans to the State governments for financing rural infrastructure projects including irrigation, soil conservation and watershed management, etc. The programme has been extended on an annual basis.

As at the end of March 2003, the total corpus of funds under the trances I to VIII of the RIDF aggregated to Rs. 28,000 crore. In 1996-97, a programme called 'Accelerated Irrigation Benefit Programme' (AIBP) which was launched by the Government of India. Under this programme, Centre provides additional central assistance by way of loans to the States on matching basis for early completion of selected large irrigation and multi-purpose projects. Rs. 8, 430 crore was released under AIBP as central loan assistance to the States during 1996-97 to 2001-02.

The Ninth Plan noted that irrigated area, which constitute only 38 per cet of the total cultivated area, contributes almost 60 per cent of the food grains production in the country. This shows the contribution of Irrigation to agricultural output. The Plan noted that the target of doubling food grains production in the next 10 years would mainly depend on the availability and performance efficiency of irrigation. What is particularly important from this point of view is the possibility of increasing the water use efficiency, which, at present, is only 38 to 40 per cent for canal irrigation and about 60 per cent for groundwater irrigation schemes.[5] It is estimated that with a 10 per cent increase in water use efficiency in irrigation systems, an additional 14 million hectares area can be brought under irrigation from the existing irrigation capacities at a very moderate investment. The plan also emphasized renovation and modernization of projects, promoting private sector participation in irrigation schemes; prompting conjunctive use of surface and ground water, ensuring progressive reduction of conveyance losses in irrigation schemes; strengthening the organization of Command Area Development in order to bridge the gap between irrigation potential created and actual utilization; completion of all on going irrigation projects to yield benefits at the earliest, etc.

DEVELOPMENT OF IRRIGATION DURING FIVE-YEAR PLANS

The outstanding feature of Indian agriculture is its dependence on rainfall. Due to unequal distribution of rainfall during the year and its variation from year to year–in respect of quantity, incidence and duration. Indian agriculture is affecting severely. Hence the development of irrigation was accorded priority in the India's Five-year plans.

Considerable amount was spent for the development of irrigation in the Five-year plans[6]. Irrigation accounted for an expenditure of Rs. 456 crores during the first Five-year plan and the irrigation potential created was of the order of 3.66 million hectares.

The expenditure on major, medium and minor irrigation projects during second and third Five-year plans was Rs. 522 crores and Rs. 909 crore respectively. An irrigation potential of 2.83 million hectares and 4.52 million hectares has been created during the second and third Five-year plans respectively.

The expenditure in annual plans (1966-69) and Fourth Five Year Plan (1969-74) accounted for Rs. 760 crore and 1,750 crore respectively. A total irrigation potential of 3.49 million hectares and 7.10 million hectares has been created in the annual plans and the Fourth Five Year Plans.

The expenditure in the Fifth Five Year Plan was Rs. 3, 073 crore and 7.92 million hectares of irrigation potential has been created. An account of Rs. 2,553 crore has been spent on irrigation in the annual plans of 1978-79 and 1979-80. A potential of 4.48 million hectares was created in the annual plans of 1978-1980. During the Sixth Five Year Plan, the total expenditure on irrigation was Rs. 9,318 crore and a potential of 11.30 million hectares has been created in the Sixth Plan.

During the Seventh Five Year Plan, the total expenditure on irrigation was 14,360 crore and the potential of 13 million hectares has been created in the Seventh Plan.

The outlay on major, medium and minor irrigation projects during Eighth and Ninth Five Year Plans was Rs. 36, 649 crore and Rs. 63,682 crore (including flood control) crore respectively. An irrigation potential of 95.40 million hectares has been created during 2000-2001. The outlay on irrigation and flood control in the Tenth Plan (2002-2007) has been kept at Rs. 1,03,315 crore.

TABLE 3.5

IRRIGATION POTENTIAL CREATED AND UTILISED UP TO THE END OF 2000-2001

(Million hectares)

Source	*Potential created*	*Potential utilized*	*Per cent utilisation*
Major and Medium Irrigation	35.3	30,5	86
Minor Irrigation	59.4	54.2	91
Total Irrigation	94.7	84.7	89

Source: Planning Commission, Ninth Five-Year Plan (1997-2002) Vol. II and Economic Survey-2001-2002

It has been estimated that India has around 166 million hectares of agricultural land out of which 140 million hectares of land would ultimately be irrigated. By 1999-2000 only 95 million hectares are provided irrigation facilities. That leaves still 45 million hectares of potential irrigation for the successive five year plans.[7] According to the Irrigation Commission, the cost of exploiting this potential would workout at Rs. 30, 000 crores (at 1980-81 prices); it would cost more than Rs. 75,000 crores now (in current prices). As the financial resources available to the Government are limited, the Seventh and Eighth Plans took up only a few major and medium projects but laid emphasis on the completion of ongoing projects.

In order to expand irrigation facilities, the Government of India has adopted the following programmes.

(a) NABARD has set up Rural Infrastructure Development Fund (RIDF) under which loans are granted to States for speedy completion of minor irrigation projects.

(b) The Government of India has launched, since 1996-97, a programme called "Accelerated Irrigation Benefit Programme" (AIBP) under which the Centre is providing additional Central assistance by way of loans to the States for early completion of selected large irrigation and multi-purpose projects. AIBP was specifically started because a large number of river valley projects-both major and medium—have spilled from plan to plan, mainly because o financial constraints of state government.

Some of these projects were in an advanced stage of construction and could provide irrigation benefits in four or five agricultural seasons. The completion of these projects, however, was beyond the resources capability of the State Governments.

Under AIBP, funds were provided to the States in the form of Centre loan assistance. The Central Government also introduced in February 2002 a Fast Track Programme for completion of some irrigation works in two or three working seasons with full Central assistance. From April 2002, AIBP has been further modified, with the adoption of a grant component, i.e. 30 per cent grant for General Category States and 90 per cent grant for Special Category States.

Between 1997 and 2004, the Government of India has released Rs. 14,670 crores as loan assistance under AIBP for 181 irrigation works and 3,810 minor irrigation works. So far 28 projects have been completed and irrigation potential of 2.2 million hectares has been created up to March 2003.

IRRIGATION PROGRAMMES IN THE TENTH PLAN

The Working Group for the Tenth Plan has assessed that total of 159 major and 242 medium irrigation projects would spill over into the

Tenth Plan (from the previous plans). Another 89 projects-known as Extension, Renovation and Modernisation (ERM) projects would also spill over to the Tenth Plan.

A further 67 major, 130 medium and 34 ERM projects would also be taken up during the Tenth Plan. For completing the old projects and executing new projects, the Tenth Plan proposes a total outlay of Rs. 1,09,025 crores and hopes to create over 11 million hectares of additional irrigation potential.

By 2005-06, the country's created irrigation potential is estimated to have gone up to about 98.1 million hectares. It is anticipated that the cumulative total of created irrigation potential by using all types of sources would go up to almost 100 million hectares by the end of Tenth Plan (2002-2007).

DECLARATION OF YEAR 2007 AS WATER YEAR

The National Common Minimum Programme assigns priority to the water sector and water management. Irrigation and drinking water are also two important components of Bharat Nirman. With a view to generating the required response from all stakeholders and to involve them in drawing up more appropriate strategies, the Government of India has declared the Year 2007 as "Water Year" with particular focus on campaigns on conservation of water through electronic and print media, mass awareness programmes by CGWB to promote rain-water harvesting and artificial recharge of ground water, and also to promote awareness on the subject through participation in festivals and fairs.

The World Water Day was celebrated on 22nd March 2007 and was inaugurated by the Hon'ble Prime Minister of India. Regional offices under the Ministry also celebrated the World Water Day. The Hon'ble President of India on 12.9.2007 gave Bhoomijal Samvardhan Puraskars and one national award for Pani Panchayat at a befitting function at Vigyan Bhavan.

DEVELOPMENT OF MINOR IRRIGATION IN STATE

Since the advent of planning, the performance of agriculture in Karnataka has been commendable. The agricultural production has been increasing at an annual rate of about 3 per cent compared to the 2.5 per cent for the country as a whole. The State is endowed with rich and varied soils and agro-climatic conditions conducive for diversified agricultural production. The full potential of agriculture could not be exploited as the State had relatively lower per centage of irrigation compared to other States.

In spite of huge investment on irrigation over the last 35 years, the state could hardly provide irrigation to one quarter of its sown area. There is hardly any scope for increasing the net sown area, which has fluctuated between 98 and 106 lakh hectares all these years, depending upon the rainfall. Evidently there are vast tracts of marginal and sub-marginal lands, which are left out of cultivation at the slightest symptoms of drought. Growth of agriculture in Karnataka, therefore, depends almost completely on intensive cultivation. It is in this context that irrigation becomes the more crucial element in agricultural development of the State.

The State usually experiences concentration of rain during the monsoon months of June to September. But rainfall becomes irregular, uneven and inadequate during this period in most of the years leading to drought situation. Therefore, major function of irrigation has been to mitigate the dangers of this calamity.

There are mainly two systems of irrigation;

(i) Surface water irrigation through river-dams, diversion weirs and storage tanks by impounding the run off and

(ii) Ground water irrigation by exploiting the sub-soil water through dug wells and tube wells.

Karnataka still lags much behind the all-India average in spite of the State 's continued efforts to increase irrigation coverage. However, since agriculture is the main sector contributing most to the State's income, Government has taken a number of steps to develop hither to undeveloped irrigation potential in case of major, medium and minor irrigation projects. As a result, expenditure as well as irrigation potential has increased over plan periods.

At the time of the state's reorganization, irrigation was available for about 8 per cent of the sown area. Development of irrigation was rightly accorded the highest priority, under planned development. Nearly one quarter of the state plan outlay has been earmarked for irrigation. The bulk of public outlay is spent on development of surface water resources.

Considerable amount was spent for the development of irrigation in Karnataka during the Five Year Plans. The expenditure on major, medium and minor irrigation projects during first, second and third five year plans was Rs, 41; 32 crores, Rs. 34.90 crores and Rs. 74.54 crores respectively. The expenditure in Annual plans and Fourth Five Year plan accounted for Rs. 50.80 crores and Rs. 170.03 crores respectively. The expenditure in the fifth plan was Rs. 230.67 crores, Rs. 664.14 crores in the Sixth plan, Rs. 773.58 crores in the Seventh plan, Rs. 2384.00 in the Eighth plan, and Rs. 6170.00 crores in the Ninth Five Year plan.

The expenditure on major, medium, and minor irrigation projects during the 2003-2004 was Rs. 3013.73 crores and 30.91 lakh hectares of irrigation potential has been created.

The State Government established the Cauvery Neeravari Nigama for better monitoring of revenue and the projects coming under the Cauvery basin at an estimated cost of Rs. 250 crores. The cumulative expenditure on the Upper Krishna Project, up to the end of November 2003 is Rs. 1507.09 crores. The responsibility of implementation of the Upper Krishna Project has been entrusted to the Krishna Bhagya Jala Nigam Limited (KBJNL). The Company has raised Rs. 110.50 crores up to the end of December 2003 through market borrowings.

The details of public investment in irrigation and trends in irrigation in Karnataka are presented in Table 3.6 and Table 3.7.

Revised estimates of expenditure on major and medium irrigation projects were Rs. 2474.27 crores under the Plan and Rs. 215.69 crores under projects pending approval in 2002-03. The expenditure anticipated in 2003-04 is Rs. 2752.91 crores for Plan projects and Rs. 50.00 crores for projects pending approval. In the current years Projects works amounting to Rs. 250 crores are being taken up through Cauvery Neeravari Nigam. Expenditure incurred up to the end of November 2003 was Rs. 1789.98 crores for Plan projects and Rs. 73.96 crores for projects pending approval. The expenditure on UKP alone up to the end of November 2003 is Rs. 1507.09 crores. Expenditure on flood control and anti-sea erosion projects in the current year is anticipated at Rs. 6.00 crores.

TABLE 3.6

PUBLIC INVESTMENT ON IRRIGATION IN KARNATAKA

(Rupees crores)

Year	*Major and Medium projects**	*Minor Irrigation works***	*Total*
1991-92	354.35	25.70	380.05
1992-93	480.17	51.52	531.69
993-94	655.68	61.75	717.43
1994-95	684.63	70.25	754.88
1995-96	920.13	61.76	981.89
1996-97	1361.12	71.57	1432.69
1997-98	1604.52	67.89	1672.41
1998-99	1719.80	89.17	1808.97
1999-2000	2027.27	107.92	2135.19
2000-01	2595.09	121.18	2716.27
2004-05	3531.24	203.96	3735.20
2005-06	3253.13	192.84	3445.97
2006-07	3764.82	445.39	4210.21

SA; Anticipated RE: Revised estimate

* Includes outlay on irrigation projects pending approval

** Surface water only

Source: Economic Survey 2006-2007, Irrigation Department Govt. of Karnataka.

TABLE 3.7

TRENDS IN IRRIGATION IN KARNATAKA CUMULATIVE POTENTIAL CREATED

(Lakh hectares)

End of plan/year	*Major and Medium projects*	*Minor Irrigation (surface)*	*Total*
VII Plan (1985-86)			
1985-86	12.27	8.59	20.86
1986-87	12.75	8.65	21.40
1987-88	12.87	8.71	21.58
1988-89	12.97	8.82	21.79
1989-90	13.09	8.92	22.01
Annual Plan			
1990-91	13.36	8.95	22.31
1991-92	13.80	9.00	22.80
VIII Plan			
1992-93	14.24	9.06	23.30
1993-94	14.94	9.13	24.07
1994-95	15.28	9.18	24.26
1995-96	15.77	9.25	25.02
1996-97	16.13	9.30	25.43
IX Plan			
1997-98	16.58	9.35	25.93
1998-99	16.93	9.38	26.31
1999-00	17.41	9.43	26.84
2000-01	18.12	9.51	27.63
2001-02	19.05	9.58	28.63
X Plan			
2005-06	21.97	9.82	31.79
2006-07	23.97	10.06	33.85

A: Anticipated, Source: Economic Survey-2006-2007.

In 2004-05 the revised expenditure on minor irrigation works was Rs. 18311.88 lakhs. In 2005-06 an allocation of Rs. 457.56 lakhs has been made for construction of new tanks Rs. 3410.58 lakhs for construction and improvements to anecuts, pickups and feeder channels. Rs. 1023.18 lakhs has been provided for Lift Irrigation schemes and Rs. 2263.50 lakhs for SCP and TSP. Besides this allocation of Rs. 85.13 lakhs has been provided for the *Zillah Panchayats* to take up minor irrigation with an atchkat of 40 hectares or below, and create a potential of 8000 hectares.[8]

The development of surface irrigation from 1993-94 onwards is summarized in Table 3.8.

TABLE 3.8

Development of Minor Irrigation from Surface Water: 1993-94 to 2006-2007

Year	*Expenditure (Rs lakh)*	*Potential created (in hectares)*
1993-94	6174.94	7000
1994-95	7025.00	5310
1995-96	6175.70	4020
1996-97	7157.33	5105
1997-98	6789.45	4838
1998-99	8917.31	3258
1999-00	10791.90	5317
2000-01	12117.71	7136
2004-05	20395.97	4280
2005-06	19284.16	7110
2006-07	44539.23	9000

RE: Revised Estimate

A: Anticipated

Source: Economic Survey of Karnataka 2006-2007, Annual Plan 2006-07

The government has formed Karnataka *Jala Samvardhane Yojana Sangha* (JSYS) within the framework of Registration of Societies Act, 1860. The main purposes of this Sangha are to develop and strengthen water sector (minor irrigation) with participatory system in tanks and ground water for improving livelihood of the rural people. It will promote and organize activities of capacity building, provide strategic resource and logistic support for training ad orientation, promote and undertake efforts for integrated, multi-dimensional interventions in related sectors and ensure timely ad adequate flow of funds from all possible sources, including World Bank.

A proposal has been approved by the World Bank to restore tanks. The budgetary allocation for JSYS during the year 2003-04 was Rs. 113.50 crores and an expenditure of Rs. 6.23 crores was incurred in 2003-04, up to the end of October 2003. JSYS has initiated series of activities as per the project implementation plan. Project partners are fully involved at various levels in ensuring organizing of capacity building programmes, review meeting at village, cluster, district and State levels, interactions with *Panchayat Raj* Institutions and line departments ad implementation of Integrated Tank Development Programmes.

SOURCES OF IRRIGATION

Important sources of irrigation in the state are canals, wells, tanks, tube/bore wells. There were 35133 tanks, 4,74,174 wells, 2,58,161 tube/ bore wells and 64 reservoirs in the state during 2000-01. The number of tanks was highest in Shimoga district with 6,499 tanks and lowest in Bidar district with only 79 tanks. Maximum number of wells was reported in Kolar district (59,608 wells or

12.4%) followed by Bijapur district (55,355 wells). Tube/bore wells were maximum in Bangalore (Rural) district (44,835) followed by Tumkur district (37,041).

AREA IRRIGATED BY DIFFERENT SOURCES

In Karnataka totally 32 Lakh hectares of gross cropped area was irrigated, out of 116,96,181 hectares of gross cropped area sown during 1999-2000.[9] The percentage of gross cropped area irrigated was 26.47 during 2002-2003 as against 22.09 during 1990-1991. The net area irrigated in the state was 23,62,661 hectares out of 1,00,74,775 hectares of net area sown during 1998-99. About 39 major food and non-food crops are grown in the state.

The gross cropped area under food crops was 79,80,89 hectares (68.2%) and that of non-food crops was 37,15,290 hectares (31.8%). Food crops are grown on nearly 7804 (in 000 ha) hectares during 2001-02. The cumulative irrigation potential under major, medium and minor irrigation (surface water) is anticipated to go up to 32.34 lakh hectares (excluding ground water) in 2005-06 from 30.92 lakh hectares in 2004-05.[10] Source wise net area irrigated during 2003-04 is presented in Table 3.9 and 3.10.

According to Table 3.8 and 3.9, in the State, out of 2384037 hectares of net area irrigated, 743383 hectares was irrigated by canals, 391329 hectares by wells, 774117 hectares by tube/bore wells, 147068 hectares by Tanks, 116656 hectares by lift irrigation and the remaining 211484 hectares by other sources. Details of area irrigated by different sources are given below.

TABLE 3.9

NET AREA IRRIGATED 2003-2004

(Hectares)

Sl. No.	District	Canals	Tanks	Wells
1	Bangalore	-	2640	801
2	Bangalore (R)	6590	15198	13844
3	Chitradurga	4627	18	-
4	Dvanagere	68542	917	2101
5	Kolar	-	371	2656
6	Shimoga	40937	46701	6028
7	Tumkur	-	7497	4583
8	Chikmagalur	5196	6755	279
9	Dakshina Kannada	-	-	45112
10	Udupi	-	1887	15890
11	Hassan	20052	16898	1459
12	Kodagu	2234	98	18
13	Mandya	68099	9029	11641
14	Mysore	86000	10888	10050
15	Chamarajanagar	8394	3172	2440
16	Belgaum	32871	379	94621
17	Bijapur	23492	164	52474
18	Bagalkot	15605	-	26658
19	Dharwad	23169	64	-
20	Gadag	20016	1246	3287
21	Haveri	2136	10579	205
22	Uttara Kannada	-	5107	5151
23	Bellary	69942	2310	6569
24	Bidar	50	694	26629
25	Gulbarga	113368	2836	26391
26	Raichur	94655	1170	18615
27	Koppal	37408	450	13827
	Total	**743383**	**147068**	**391329**

Source: ASCR 2003-04, Karnataka at a Glance 2004-2005 p. 74

TABLE 3.10

Net area irrigated 2003-2004

(Hectares)

Sl. No.	*District*	*Tube Wells*	*Lift Irrigation*	*Other Sources*	*Total*
1	Bangalore	9676	-	-	13117
2	Bangalore (R)	34188	900	-	70720
3	Chitradurga	64402	-	-	69047
4	Dvanagere	44799	7669	644	124672
5	Kolar	69116	-	-	72143
6	Shimoga	12255	5188	7007	118116
7	Tumkur	105803	-	-	117883
8	Chikmagalur	7099	142	7772	27243
9	Dakshina Kannada	9045	2261	16980	73398
10	Udupi	460	5227	8903	32397
11	Hassan	32732	123	5198	76462
12	Kodagu	10	85	705	3150
13	Madya	6348	448	1176	96741
14	Mysore	7850	375	-	115163
15	Chamarajanagar	31037	176	-	45219
16	Belgaum	65313	54701	49942	297827
17	Bijapur	35424	-	17036	128590
18	Bagalkot	46065	1738	67520	157586
19	Dharwad	13784	-	115	37132
20	Gadag	26856	6115	5246	62766
21	Haveri	23995	11946	2141	51002
22	Uttara Kannada	1203	385	9193	21039
23	Bellary	51491	10202	6954	147468
24	Bidar	11886	102	1137	40498
25	Gulbarga	4332	886	3815	151628
26	Raichur	9231	7557	-	131228
27	Koppal	49717	400	-	101802
	Total	**774117**	**116656**	**211484**	**2384037**

Source: ASCR 2003-04, Karnataka at a Glance 2004-2005 p. 74

Canals: Important canals in the State are Cauvery, Tungabhadra, Upper Krishna, Malaprabha, and Bhadra projects. About 11,83,434 hectares of gross cropped area was irrigated by canals during 1997-98 as against 11,92,542 hectares during 1996-97. District wise, major coverage of gross cropped cultivated area by canal irrigation was in Raichur (2.3 lakh hectares) followed by Bellary (1.4 lakh hectares) and Mysore (1.2) lakh hectares). Belgum, Bellary, Gulbarga, Mandya, Mysore, Raichur and Shimoga districts together cover about 76% (9,08,627 hectares) in the State during 1997-98. In 2003-2004 the net area irrigated through canals was 743383 hectares as against 9,18,343 hectares during 1996-97.[11] Among the districts Gulbarga had the largest net area irrigated (113368 hectares) followed by Raichur (94655 hectares).

Tanks: Before the formation of projects, tank irrigation was popular in the State. During 1997-98, the gross cropped area irrigated through tanks was 2,66,886 hectares as against 2,72,239 hectares during 1996-97. Among the districts, largest gross cropped area irrigated was in Shimoga (73728 hectares) followed by Hassan (31,440 hectares) and lowest was in Kodagu (only 122 hectares). In 2003-2004 the net area irrigated in the State was 147068 hectares as against 2,40,361 hectares during 1996-97. District wise, highest net area irrigated was in Shimoga (46701 hectares) followed by Hassan (16898 hectares) and Bangalore Rural (15198 hectares).

Wells: Gross cropped area irrigated through wells was 5,43,461 hectares during 1997-98 as against 5,67,262 hectares during 1996-97. Among the districts largest gross cropped area irrigated was reported in Bijapur (133816 hectares), which constituted 24.6 % of the gross area irrigated through wells in the State. In 2003-2004 net area irrigated through wells was 391329 hectares as against 4,65,562 hectares during 1996-97. Among the districts Belgaum had the largest net area irrigated (94621 hectares) followed by Bijapur (52474 hectares).

Tube/Bore Wells: Gross area irrigated through tube/ bore well was 5,14,650 hectares during 1997-98 as against 4,70,747 hectares during 1996-97. Among the districts, largest gross area irrigated through tube/bore well was in Chitradurga (73,184 hectares) followed by Tumkur (69,873 hectares). Net area irrigated through tube/bore wells 774117 hectares in 2003-2004 as against 367268 hectares during 1996-97. Among the districts, largest net area irrigated was reported in Tumkur district (105803 hectares) followed by Kolar (69116 hectares).

Other Sources: In the state, 328140 hectares of net area was irrigated through other sources, which includes 116656 net area irrigated through lift irrigation.

During 2005, the quantum of rainfall received from the North-East monsoon was 246 mm as against 195 mm of normal rainfall and 122 *taluks* had normal to excess rainfall while 48 *taluks* faced deficit and 5 suffered from scanty rainfall.

Thus, the major portion of Karnataka's land remains thirsty for want of irrigation facilities. In view of the importance of water for agriculture, emphasis on minor irrigation has become an issue of inescapable importance in Karnataka.

REFERENCES

1. Report of the Royal Commission on Agriculture, p. 268.
2. Economic Survey, 2006-07, p. 170.
3. CSO Statistical Abstract, 2002.

4. Government of India, Tenth Five Year Plan, 2002-07 (Delhi, 2003), VolumeII, Annexure 8.1.2, p .892
5. Government of India Ninth Five Year Plan 1997-2002 (Delhi 1999), Volume II, p. 482.
6. Planning commission Ninth Five-year plan (1997-2002) Vol. II.
7. Planning Commission, Ninth Five-Year Plan (1997-2002) Vol. iI and Economic Survey-2001-2002.
8. Annual Plan 2004-2005, Government of Karnataka.
9. *Karnataka at a Glance*, Economic Survey – 2003-2004.
10. *Economic Survey of Karnataka*, 2005-06.
11. *Karnatakada Arthashastra*, Ne. Thi. Somashekar, p. 259.

4

IMPACT OF TUBE WELL IRRIGATION ON AGRICULTURE

The strategic role of irrigation as an essential input for agriculture hardly needs any emphasis. Irrigation possesses significant 'area effect, 'crop pattern effect' and 'yield effect'. It increases the net area cultivated and more importantly the 'gross cropped area' by enhancing the crop intensity through double or multiple cropping. Moreover, the favourable impact of irrigation on creation of additional rural employment for agriculture labourers and self-employed family workers of cultivating households is being highlighted in Karnataka. Irrigation reduces the rigours of unemployment, under employment and seasonal unemployment by enhancing the mandays of employment in an abundantly labour surplus rural economy.[1] In this background the study is undertaken.

Before we undertake the impact of tube well irrigation on agricultural production with special reference to paddy cultivation, it would be useful to study the general features of Hassan district. It is one of the predominant districts in terms of area under irrigation, particularly tube well irrigation. Hence, research on irrigation in this district is considered important. The present chapter proposes to study the pattern of land holdings, land utilization, climate and rainfall, soils, irrigation, cropping pattern and impact of tube well irrigation on agricultural production in the study area.

AGRICULTURAL POPULATION

The economy of the district is primarily based on agriculture, which is the chief occupation of the people as in other contiguous districts. According to the 2001 Census, the total population of the district was 17,21,669. This is broadly classified as 8,64,711 workers and 8,56,958 non-workers, the workers being further classified under nine categories according to their economic activities. So far as this chapter is concerned, the first two categories of workers, namely, cultivators and agricultural labourers are taken into consideration. The total number of persons under these two categories was 6,02,475. The Table 4.1 shows the distribution of agricultural workers by taluks, based on the 2001 Census.[2]

The following table shows that the proportion of agricultural workers or cultivators to the total working force was highest in Channarayapatna taluk followed by Hassan and Arsikere taluks. But this proportion was remarkably low in Sakleshpur taluk, due to considerable number of people are engaged in Coffee plantation. But the largest percentage of agricultural labourers was found in Arsikere taluk, followed by H.N. Pura, Arkalgud and Hassan taluks.

TABLE 4.1

CLASSIFICATION OF AGRICULTURAL WORKERS–2001

Sl. No.	Taluk	Cultivators			Agricultural Labourers		
		Male	*Female*	*Total*	*Male*	*Female*	*Total*
1	Alur	12440	6265	18705	2406	4274	6680
2	Arkalgud	41161	24224	65385	6804	11881	18685
3	Arsikere	54624	29916	84540	10512	18542	29054
4	Belur	30962	17216	48178	5614	9128	14742
5	C.R. Patna	59618	43046	102664	5484	10025	15509
6	Hassan	50834	42481	93315	4701	10995	15696
7	H. N. Pura	33146	18417	51563	6131	14241	20372
8	Sakleshpur	7039	4427	11466	2451	3470	5921
	To tal	28982	185992	475816	44103	82556	126659

Source: 2001 Census, Hassan District At a Glance-2005-2006, District Statistical Office, Hassan

TABLE 4.2

AGRICULTURAL LANDHOLDING - 2001

Sl No.	Taluk	Marginal (Below 1He.)	Small (1-2 He)	Semi medium (2-4 He)	Medium (4-10 He.)	Large (More than 10 He.)	Total
		Number	Number	Number	Number	Number	Number
1	Alur	14676	5749	2365	852	166	23810
2	Arkalgud	31985	11358	4836	1129	69	49377
3	Arsikere	42993	18764	8186	2186	127	72256
4	Belur	27025	11838	4643	1712	300	45518
5	C.R. Patna	64812	19739	7263	1573	54	93441
6	Hassan	61134	14407	4771	869	67	81248
7	H. N. Pura	33402	9729	3353	727	29	47240
8	Sakleshpur	12200	4730	3206	2314	669	23119
	Total	288229	96314	38623	11362	1481	436009

Source; Agriculture Census 2001, Hassan District At a Glance-2005-2006, District Statistical Office, Hassan.

PATTERN OF LAND HOLDINGS

As per the Agricultural Census of 2001, land holdings in Hassan district are characterized by the predominance of small holdings. The district has got a proportion of 66 per cent of marginal, 22 per cent of small, 9 per cent of semi medium and 3 per cent of medium and 0.5 per cent of big landholders.[3] Thus the inequality in the distribution of land is obvious. The agricultural land holdings and area in the district is shown in the following table.

LAND UTILISATION

At the end of the year 2004-2005, the geographical area of the district was 6,62,602 hectares of which 4,41,682 hectares is used for cultivation purposes. Area under cultivation comprised, net area sown, and fallow land, which is temporarily out of cultivation for various reasons. Net area sown was 3,93,481 hectares and fallow land is 48,202 hectares, area under forests was 58,775 hectares. Barren and uncultivated land put to the non-agricultural use constitutes 108,726 hectares. The following Table 4.3 shows the land utilization pattern in the district for the year 2004-05.

TABLE 4.3

LAND UTILISATION 2004-05

Sl. No.	*Category*	*Area in Hectares*
1	Geographical Area	6,62,602
2	Forests	58,775
3	Area not available for cultivation	108,726
4	Permanent Pastures	32,943
5	Trees and Groves	6,957
6	Cultivable Waste	13,518
7	Other fallow lands	21,813
8	Current fallows	36,389
9	Net Area Sown	3,93,481

Source; Agriculture Census 2001, Hassan District At a Glance-2005-2006, District Statistical Office, Hassan.

The area under permanent pastures and other grazing lands, area under forests and tree crops and cultivable waste are more in malnad and semi-malnad parts.[4] Depending upon the seasonal conditions, the area under cultivation tends to differ from year to year. But by stepping up irrigation facilities, undertaking soil conservation measures, improvement of wastelands, etc. efforts are being made to increase the extent of land under cultivation. Sustained efforts are also being made to increase the total yield from the cultivated area by adoption of various measures, so as to make agriculture a profitable venture.

CLIMATE AND RAINFALL

The Hassan district has an agreeable climate. The year may be divided into four seasons according to the climatic conditions. The summer season is from March to the end of May and it is followed by the south-west monsoon season lasting up to about the end of September. October and November may be termed the post-monsoon or retreating monsoon season. The period from December to February is the dry season with generally clear and bright weather.

There were 58 rain-gauge stations in the district in 2006 in working condition. The average annual rainfall in the district is 1,031 millimetres. The western part of the district in the vicinity of the Western Ghats gets heavy rainfall. Most of the rainfall in the district is confined to the period from May to October. July being the rainiest month. The rainfall during the south-west monsoon months, i.e., from June to September, constitutes only about 59 per cent of the annual normal rainfall. The rest of the rainfall is received mainly during the pre-monsoon months of April and May and the post monsoon season. The rainfall during the pre-monsoon months is mainly in the form of thunder-showers.

SOILS

The soils of Hassan district show a marked diversity in different parts of the district depending upon the nature of the parent rock and the climatic conditions of the respective areas. It has forest loams and red laterite, ferruginous and clayey soil of the hill slopes and the sandy or gravelly soil of the plains. The soils of the western taluks covering Alur and Belur are derived from granites, laterites and schists. These are mostly suitable for irrigated and plantation crops like coffee, tea, pepper, cardamom, areca, paddy and sugarcane.

The soils of the eastern *taluks* comprising Hassan, Channarayapatna, Arsikere and Holenarsipur are red sandy soils, derived from granites, gneiss and schists. They are well drained but poor in bases and water holding capacity and are favourable for growing crops like paddy, sugarcane, coconuts, potato, vegetables and plantain crops under irrigated conditions and *ragi*, millets, pulses, groundnut, potato and *jowar* under rainfed conditions. Occasional patches of black soil also appear in Arsikere taluk.

But large areas of the district contain alkaline soils, nearly 65 per cent of the land having PH ranging from 8.0 to 9.0. Almost all the soils are poor in available Phosphorus, only 5 per cent of soils being sufficient in this regard.

IRRIGATION

The three major rivers flowing in the district are Cauvery, Hemavathi, and Yagachi. Hemavathi is the perennial river; the length of this river is 192 kms. As the district comprises *malnad,* semi-*malnad* and *maidan* parts, the sources and the systems of irrigation also vary. In *malnad* and semi-*malnad* areas, there are a number of small anicuts, tanks and pick-ups constructed across the rivers. In the *maidan* parts, tank irrigation is predominant. The net area irrigated in the district is shown in the Table 4.4.

From the above table it is observed that Ground water potential is more in all the taluks of district. Main source of irrigation is tube wells in the district.

The area irrigated by tube wells in the year 2006 was 34,387 hectares, followed by canals, which is 33,359 hectares, 25,797 hectares by tanks, 910 hectares by wells, 206 hectares by lift irrigation and 944 hectares by other sources. The total net area irrigated in the district by all sources is 95,603 hectares. Therefore, ground water utilization plays an important role in irrigation in Hassan district.

CROPPING PATTERN

The *malnad* parts have mostly one crop from rain fed terraced paddy fields. Cultivation in these parts is skilled but by no means intensive. The farmer is able to grow not only a reliable subsistence crop, but also cash crops like cardamom, coffee and pepper. Superimposed on the traditional pattern of farming generally prevalent in these parts. The recorded acreages of oilseeds in these parts are very low. The semi-*malnad* parts exhibit a striking contrast

TABLE 4.4

NET AREA IRRIGATED 2005-2006

(Hectares)

Sl. No	*Taluk*	*Canals*	*Tanks*	*Wells*	*Tube wells*	*Lift irrigation*	*Other sources*	*Total*
1	Alur	1471	5212	62	134	0	322	7201
2	Arkalagud	10790	4844	84	802	150	126	16796
3	Arsikere	0	0	0	11662	0	0	11662
4	Belur	905	4418	279	2692	0	496	8790
5	C. C. R. Patna	4368	756	19	14345	0	0	19488
6	Hassan	1854	5194	58	4154	0	0	11260
7	Holenarasipur	13971	941	158	270	56	0	15396
8	Sakaleshpur	0	4432	250	328	0	0	5010
	Total	33,359	25797	910	34387	206	944	95603

Source: Hassan District At a Glance-2006-2007, District Statistical Office, and Hassan.

to the *malnad* proper. Here, though rice is the most important cereal, there are appreciable acreages of *ragi* as well as smaller acreages of other cereals in Belur, Arakalagud and Hassan taluks. Coconut has a strong hold in Arsikere, while potato is associated with Hassan and channarayapatna taluks. In the *maidan* parts, double cropping forms an important feature, chiefly in irrigated tracts. The dry fields are mostly used for *ragi* growing as well as for cultivation of groundnut as a commercial crop and in some places for growing tobacco and cotton. The wetlands are used for growing paddy or sugarcane. A considerable part of the dry area in Arsikere and Channarayapatna taluks is under coconut cultivation.

In the rain fed areas, *ragi,* groundnut, hybrid maize and hybrid *jowar* followed by horse gram, potato followed by *poorna ragi.* In places where two crops are grown under tank irrigation, the pattern is different. Paddy followed by groundnut, *ragi* or hybrid maize in summer season, paddy and sugarcane is grown alternately, once in two years.

The district had 3,93,481 hectares of land under cultivation (net sown area) in 2004-05.[5] The distribution of acreages of most important crops among the various taluks of the district, as in 2005-06, are as follows:

The above table shows that, ragi and paddy are the most important cereals of the district and they are grown in all the *taluks,* while the acreage under *ragi* in Sakaleshpur *taluk* is negligible; the acreage under paddy in Arsikere *taluk* is small. Jowar is grown mainly in Arsikere taluk. While Arsikere and Channarayapatna are noted for coconut cultivation, sugarcane is grown mainly in Channarayapatna, Holenarasipura and Belur taluks. While cardamom is almost confined to Sakaleshpur taluk, Coffee is grown in Sakaleshpur, Belur and Alur taluks.

TABLE 4.5

CROP PATTERN - 2004-2005

(In Hectares)

Sl. No.	Taluk	Paddy	Ragi	Jowar	Maize	Potato	SugarCane	Coconut
1	Alur	6079	2245	-	4333	2115	8	607
2	Arkalagud	11855	12286	-	4780	8880	33	1845
3	Arsikere	357	32047	2222	2145	1565	95	26140
4	Belur	5340	8183	730	5935	7188	712	1965
5	C. C. R. patna	4670	21840	30	2945	1360	5731	23526
6	Hassan	4510	15680	90	10280	32566	508	4505
7	Holenarasipur	11191	14528	-	1802	2112	1843	3094
8	Sakaleshpur	8530	28	-	5	5	-	93
	Total	52532	106837	3072	32225	55791	8930	61775

Source: *Hassan District At a Glance-2006-2007, District Statistical Office, Hassan.*

In study area, sometimes the availability of water from rivers and reservoirs is not sufficient. Moreover, all tanks are rainfed. So the only source of irrigation, which provides perennial water supply for cultivation, is tube wells.

IMPACT OF TUBE WELL IRRIGATION

Among the sources of minor irrigation like tanks, wells and tube wells, in recent years, tube wells have become more popular and the use of tube wells as a source of minor irrigation has acquired significance in most of the taluks in Hassan district. Therefore this study covers two villages namely, Dammaningala and A. Kalenahalli of Channarayapatna taluk in Hassan district. Channarayapatna, which is one among the 8 *taluks* of the Hassan district. It constitutes 6 hobli hqs, 40 *Grama Panchayaths,* 375 inhabited villages and 10 un-inhabited villages.

The total geographical area of Channarayapatna is about 103,464 hectares. Among them 697 hectares is covered by forest, 5319 hectares by barren, 10359 hectares by non-agricultural, 870 hectares cultivable waste, 9275 hectares by permanent pasture and 374 hectares is covered by trees and groves. Fallow lands account for 12,186 hectares, the land cultivated more than once is about 3,045 hectares and net area is about 64,384 hectares.

The population of the *taluk,* according to the Census of 2001, was 2,78,105 of which 1,37,573 were males and 1,40,532 were females. In 2001 agriculture Census there is 93441 landholders and it extended to 80929 hectares.

In 2004-05, the net area irrigated in this *taluk* from all sources was 15,908 hectares. Out of this, an area of 3,825 hectares was irrigated by canals, 1,215 hectares by tanks,

12 hectares by wells and 10,856 hectares by tube wells. The average rainfall of the *taluk* is about 712 m.ms. *Ragi*, paddy, *jowar*, maize are the major food crops. Sugarcane, potato, coconut, banana, are the commercial crops of the district. When we observe this, we can understand the impact of tube well irrigation on agricultural production.

The impact of tube well irrigation on agricultural production with special reference to paddy cultivation in study area can be analysed on the basis of primary data are as follows.

TABLE 4.6

DISTRIBUTION OF THE EDUCATIONAL BACKGROUND OF HOUSEHOLD IN SELECTED VILLAGES

Fig in number

Educational Qualifications	*Total No of HH*
Up to SSLC	43
PUC	13
Degree	10
B.Ed	11
Master Degree and Above	3
ITI	1
Diploma	6
Nil Education	13
Total	100

Source: *Field Data*

The table 4.6 assesses the educational qualification of the head of the household. Out of the total 100 household surveyed 43 individuals has studied up to SSLC, 13 have studied up to PUC, 10 have completed degree and 11 have done B.Ed. Some of the individuals have qualified technical cources also. About 6 have diploma and one

ITI, 3 of the individuals have master degree of the total, 13 individuals are not having any educational background.

The Table 4.7 explains the total family members and their dependency on agriculture. From this table, we can understand that the households with four members in the family were maximum. About 32 households have four members in their family. The cost is one household with nine members in this family. It is observed that the members dependent on agriculture were maximum in the household having four members in their family. It was least in the household having nine members in the family.

The mean is also high of the households having four members and least for households having nine members. The mean value for households having four members is 173.33 and for that of household having nine members is 12.00.

The co-efficient of variation for 2 members household, 3, 4, 5, 6, 7, 8 and 9 members household is 52. 03 per cent, 45.07 per cent, 43.32 per cent, 43.63 per cent, 48.80 per cent, 43.30 per cent, 47.19 per cent and 43.30 per cent.

The table 4.8 indicates the ownership of irrigated and other lands, which has tube wells and open wells of the total irrigated (owned) land 257 acres in irrigated using tube wells and 31 acres irrigated using open wells. About one acre of leased land in irrigated land using tube well water and two acres is irrigated by open well. Tube wells are used to irrigate 4 acres of irrigated leased out land. There is no leased out land, which is using open well water for irrigation. About 40 acres of land other than irrigated land is irrigated using tube wells. The Mean for

TABLE 4.7

TREND AND VARIATION IN TOTAL FAMILY MEMBERS OF SELECTED VILLAGES

Fig in number

Total Family Members	2	3	4	5	6	7	8	9	*Total*
No of HH.	(4) 8	(15) 45	(32) 128	(13) 65	(26) 156	(4) 28	(5) 40	(1) 9	(100) 475
Dependents of Agri.	(9) 18	(21) 63	(33) 132	(15) 75	(14) 84	(04) 28	(03) 24	(01) 9	(100) 433
Total	(13) 26	(36) 108	(65) 260	(25) 140	(40) 240	(8) 56	(8) 64	(2) 18	(200) 912
Mean	17.33	72.00	173.33	93.33	160.00	37.33	42.67	12.00	-
SD	9.02	32.45	75.08	40.72	78.08	16.17	20.13	5.20	-
CV	52.03	45.07	43.32	43.63	48.80	43.30	47.19	43.30	-

Source: *Field Data* Bracate indicates the total number of households.

irrigated (owned), irrigated (leased), irrigated (leased out) and other is 144 acres, 1, 5, 2 and 20 acres respectively.

TABLE 4.8

TREND AND VARIATION IN LAND PARTICULARS OF SELECTED VILLAGES

Fig in acres

Particulars	*Irrigated (Owned)*	*Irrigated (Leased in)*	*Irrigated (Leased out)*	*Others*
Tube Well	257	1	4	40
Well (Open)	31	2	0	0
Total	288	3	4	40
Mean	144	1.5	2	20
SD	159.81	0.71	2.83	28.28
CV	110.98	47.14	141.42	141.42

Source: Field Data.

Co-efficient of variation for irrigated (owned) land is 110.98 per cent irrigated (leased in) is 47.14 per cent irrigated (leased out) is 141.42 per cent and others is 141.42 per cent.

The table 4.9 depicts the net sown area in the irrigated and unirrigated land and it is 313 and 295 acres respectively. The total irrigated land, which grows paddy, is 152 acres. Only 8 acres of unirrigated land was used for paddy cultivation. About 287 acres of irrigated land is used to grow other crops. Irrigated land used to grow crops other than paddy is about 161 acres. About 20 acres of irrigated area were not used for cultivation, compared to 8 acres of unirrigated land. Mean land utilised for different agricultural activities in irrigated area is 156.5, and in unirrigated it is 147.5.

TABLE 4.9

TREND AND VARIATION IN LAND CULTIVATION IN SELECTED VILLAGES

Fig in acres

Item	*Irrigated*	*Unirrigated*
Net Sown Area	313	295
Area for Paddy	152	8
Area Under Other Crops	161	287
Area not Cultivated	20	8
Total	646	598
Mean	156.5	147.5
SD	6.36	197.28
CV	4.07	133.75

Source: Field Data

The co-efficient of variation for irrigated land is 4.07 per cent and for that of unirrigated land is 133.75 per cent.

TABLE 4.10

TREND AND VARIATION IN MATERIAL INPUTS OF SELECTED VILLAGES

Fig in Rupees

Particulars	*Owned*	*Purchased*
Seeds	21320	26050
Organic Manure	63290	24610
Fertilizer	260	54050
Pesticides	800	37980
Total	85670	142690
Mean	21417.50	35672.50
SD	29586.24	13638.53
CV	138.14	38.23

Source: Field Data

The Table 4.10 discusses the cost incurred by the 100 households. The respondents produced and owned seeds worth of Rs. 21,320 to be sown in the field and in addition spent Rs. 26,050 for purchasing it from market. The cost of organic manure owned was about Rs. 63,290 and purchased was Rs. 24,610. The fertilizer that was stocked at home was worth Rs. 260 and was also purchased for Rs. 54,050. Pesticides were purchased for Rs. 37,980, in addition to the pesticides owned worth of Rs. 800. Mean of Rs 35,672.50 was spend to purchase materials in addition to the materials owned which is worth of Rs. 21,417.50. Co-efficient of variation for the owned material inputs was 138.14 per cent and for that of purchased material was 38.23 per cent.

TABLE 4.11

TREND AND VARIATION IN FARM LABOUR OF SELECTED VILLAGES

Particulars	*Family Labour*	*Hired Labour*	*Total(In Rs.)*
Preparation For Nursery	270	238	508
Transportation			&
Applications of Manure	230	300	530
Preparations	For		Beds
Ploughing etc.	251	465	716
Transplantation	254	713	967
Operations in respects of Crops Standing weeding etc	257	541	798
Harvesting and Threshing	258	809	1067
Other Operations	263	160	423
Total	1783	3226	5009
Mean	254.71	460.86	715.57
SD	12.54	244.02	243.80
CV	4.92	52.95	34.07

Source: Field Data

The Table 4.11 reflects the number of individuals working as family labour and hired labour in various farm activities. Among family labour, around 250 individuals were involved in different types of labour. Number of hired labour used for majority of the farm works was more than family labour. Preparations for nursery and other operation used less hired labour than family labour. The mean of family labour and hired labour is 254.71 and 460.86 respectively. It is observed that co-efficient of variation for family labour is 4.92 per cent and for that of hired labour is 52.95 per cent.

TABLE 4.12

TREND AND VARIATION IN ANIMAL LABOUR AND TRACTOR LABOUR OF SELECTED VILLAGES

Fig in Rupees

Particulars	*Animal Labour*	*Tractor Labour*
Ploughing leveling	37930	32350
Transportation of Manure	13840	30652
Transportation of main Products By Products	17400	50802
Total	69170	113804
Mean	23056.67	37934.67
SD	1300309.32	1117573.27
CV	5639.62	2946.05

Source: Field Data

The Table 4.12 enumerates the cost met for animal and tractor power during different agricultural activities.

Animal labour costs Rs. 37,930 for ploughing, while tractor power costs Rs. 32,350. The cost of tractor power is high for transportation of manure and main products/

byproducts. If costed Rs. 30,652 and Rs. 50802 respectively. The animal labour used for transporting manure and main products/byproducts was Rs. 13,840 and Rs. 17,400. The mean cost for animal labour and tractor power is Rs. 23,056.67 and Rs. 37,934.67 respectively.

Co-efficient of variation for animal labour is 5639.62 per cent and for that of tractor power is 2946.05 per cent.

TABLE 4.13

DISTRIBUTION OF THE OTHER EXPENDITURE IN DIFFERENT AGRICULTURAL ACTIVITIES OF SELECTED VILLAGES

Particulars	*Amount (Rupees)*
Irrigation	45140
Land Revenue	21250
Interest on Crop Loans	7850
Repairs and Maintenance of Agriculture Implements and Machinery	15200
Spraying and dusting	18400
Others	300
Total	108140

Source: Field Data

The Table 4.13 indicates the total expenditure for different activities. The total cost incurred for irrigation was more than any other activity. The total cost spend on irrigation is Rs. 45,140. Least money was spend on other activities, i.e. only Rs. 300. Paid towards interest on crop loan was Rs. 7,850; for repairs and maintenance of agricultural implements and machinery was Rs. 15,200 and for spraying and dusting was Rs. 18400.

TABLE 4.14

INVENTORY AND FARM ASSETS OF SELECTED VILLAGES

Particulars	*Value in Rupees*
Land and Buildings	93750000
Irrigation Sources	5250000
Agricultural Implements	15000000
Live Stock	750000
Total	114750000

Source: Field Data

The Table 4.14 represents the monetary values of inventory and farm assets. Land and building were valued at Rs. 93750000 and it was highest compared to values of irrigation sources, agricultural implements and livestock. The total value of livestock as assets was Rs. 750000, They have Rs. 15000000 worth agricultural implements and Rs. 5250000 worth irrigation sources.

TABLE 4.15

FARM INCOME OF SELECTED VILLAGES

Particulars	*Income in Rupees*
Main Product	2293850
By Products	540800
Total	2834650

Source: Field Data

This Table 4.15 represents the farm income earned from main products and byproducts. The farm income derived from main product is Rs. 22,93850 and the income derived from byproducts is Rs. 5,40800. The income derived from main product was higher than that of income derived from its byproducts.

TABLE 4.16

TREND AND VARIATION IN INCOME FROM OTHER SOURCES OF SELECTED VILLAGES.

Fig in Rupees

Particulars	*Produced*	*Sold*
Milk and Milk Products	68000	42000
Other Farm Produce such as Timber, Manure etc.	125300	100000
Sale of Live Stock and Poultry	230000	210000
Rent and Share	10000	10000
Income from fixed Assets	9080950	9080950
Rent from Tractor/Tiller	2500000	2500000
Others	120000	120000
Total	12134250	12062950
Mean	1733464.29	1723278.57
SD	3360731.27	3366407.42
CV	193.87	195.35

Source: Field Data

The Table 4.16 assesses the income from other sources (production and quantity sold). The production of milk and milk products was worth Rs. 68,000; other farm products such as timber manure etc. was Rs. 125300 sale of livestock and poultry was Rs. 2,30,000; Rent and Share was Rs. 10,000; income from fixed assets was 90,80950; rent from tractor/tiller was 25,00,000 and through others was Rs. 1,20,000.

The cost of selling was same for rent and share; income from fixed assets; rent from tractor/tiller and others. The mean for produced was 17,33464.29 and for sold is 17,23,278.57.

The co-efficient of variation for produced is 193.87 per cent and for that of sold is 195.35 per cent.

Thus the field survey has revealed the above information on which conclusions are drawn and suggestions are given.

IMPACT ON RESOURCE USE EFFICIENCY

Resource use efficiency means that land, human labour, animal labour, seeds, fertilizers, organic manures, power, pesticides, public investment, subsidies and credit for agriculture are used in an efficient manner. The tube well irrigation confers a number of benefits, both direct and indirect, on the community. It paves the way for adoption of multiple cropping and use of modern inputs including improved varieties of seeds, thereby enhancing agricultural production. It stabilizes crop yield by saving crop from drought situations and it increases employment potential in rural areas through changes in the cropping pattern.

Therefore an attempt has been made to study the resource use efficiency on irrigated farms in Hassan district of Karnataka, resource use efficiency analysis implies that there is an ample scope to increase the productivity of the irrigated farmers by adopting proper technology as well as optimum allocation/utilization of the available resources. Efficiency of the farmers could be supported by tube well irrigation, the results of this has indicated that irrigated farmers are efficient than the non-irrigated farmers. This reveals that there is wider scope for further improvement in the tube well irrigation system.

The impact of tube well irrigation on resource use efficiency in paddy cultivation in study area can be analysed on the basis of primary data are as follows.

1. Utilisation of Land

The application of modern inputs to agriculture and the raising of productivity levels become possible with irrigation. The extent of irrigation facilities available to the farmer is a key to improving agricultural productivity. Table 4.1 shows that the ownership of irrigated and other lands, which has tube wells and open wells of the total irrigated (owned) land 257 acres in irrigated using tube wells and 31 acres irrigated using open wells. About one acre of leased land in irrigated land using tube well water and two acres is irrigated by open well. Tube wells are used to irrigate 4 acres of irrigated leased out land. There is no leased out land, which is using open well water for irrigation. About 40 acres of land other than irrigated land is irrigated using tube wells.

As per Tables 4.8 and 4.9, maximum number of households cultivated through tube well irrigation with owned tube wells. But the land leased in and leased out land is very least number of acres could be observed. The tube well irrigation is dominant when compared to well irrigation. As a result, in spite of the advent of tube wells, a sizable proportion of their land is under open well irrigation. But this is not so in the case of marginal and small farmers. The latter have been recently introduced to irrigation, and the chief source of irrigation for them is the tube well.

The Table 4.9 depicts the net sown area in the irrigated and unirrigated land and it is 313 and 295 acres respectively. The total irrigated land, which grows paddy, is 152 acres. Only 8 acres of unirrigated land was used for paddy cultivation. About 287 acres of irrigated land is used to grow other crops. Irrigated land used to grow crops other than paddy is about 161 acres. About 20 acres of irrigated area were not used for cultivation. It was found that unirrigated land is equal to that of irrigated land in the study area. Irrigated land has been utilized to the maximum extent for paddy cultivation. But unirrigated land is used for growing other crops. Therefore, the proportion of irrigated land under tube wells is considerably higher in the paddy cultivation.

2. Increasing Use of Material Inputs

It is examined that the changes in crop pattern in the district due to minor irrigation. Under tube well irrigation the paddy crop grown has increased. It required material inputs such as seeds, chemical fertilizers and pesticides in addition to the traditional farmyard manure and green manure. Table 4.10 indicates the use of material inputs in paddy cultivation under tube well irrigation.

According to Table 4.10, the respondents produced and owned seeds worth of Rs. 21, 320 to be sown in the field and in addition spent Rs. 26, 050 for purchasing it from market. The cost of organic manure owned was about 63,290 and purchased was Rs. 24610. The fertilizer that was stocked at home was worth Rs. 260 and was also purchased for Rs. 54050. Pesticides were purchased for Rs. 37980, in addition to the pesticides owned worth of Rs. 800.

It was observed that highest amount was spent in purchase of materials and lowest cost could be observed in case of owned materials. The number of indications (in Table 4.3), it could be observed in the case of fertilizer and pesticides registered only 260 and 800 respectively in case of materials owned. But whereas in the case of purchased materials it was found to be higher amount i.e. 54050 and 37980 when compared to owned materials.

3. Use of Human Labour and Animal Labour

In any analysis of the utilization of inputs, one of the aspects that well be of interest is the impact of irrigation on the use of labour, both human and animal. The findings of the impact of tube well irrigation on labour-use well have very useful implications for employment.

The Table 4.11 reflects the number of individuals working as family labour and hired labour in various farm activities. Among family labour, around 250 individuals were involved in different types of labour. Number of hired labour used for majority of the farm works was more than family labour. Preparations for nursery and other operation used less hired labour than family labour. The mean of family labour and hired labour is 254.71 and 460.86 respectively. It is observed that co-efficient of variation for family labour is 4.92 per cent and for that of hired labour is 52.95 per cent.

The Table 4.12 enumerates the cost met for animal and tractor power during different agricultural activities.

Animal labour costs Rs. 37,930 for ploughing, while tractor power costs Rs. 32,350. The cost of tractor power is high for transportation of manure and main products/ byproducts. If costed Rs. 30,652 and Rs. 50,802 respectively. The animal labour used for transporting manure and main products/byproducts was Rs. 13,840 and Rs. 17,400. The mean cost for animal labour and tractor power is Rs. 23,056.67 and Rs. 37,934.67 respectively.

It was found that number of hired labour used for majority of the farm labour was more than family labour. The cost of animal labour and tractor labour indicates the cost of production in main products i.e. paddy cultivation. When compared, tractor labour spending is more than the animal labour. Therefore, after the introduction of tube well irrigation system, both human labour requirements and animal labour requirement has increased in paddy cultivation.

4. Employment

While considering the individual components of the total cost, it was found wages to be a major factor in the higher cultivation cost of paddy. When viewed from the macro angle, this has obviously a social benefit. The fact that wet cultivation requires more labourers which indicates that the employment generating capacity of wet cultivation is greater. Paddy requires almost 2 to 3 times more labourers than *ragi* does. Such agricultural operations as transplantation, weeding and harvesting which are needed for paddy are labour-intensive. The cultivation of wet crop, therefore, leads to greater employment. In this sense, the impact of tube well irrigation on employment is positive and definite.

5. Increasing Income of the Household

The increased yields on the one side and the cultivation of paddy crop on the other were instrumental in raising the average income of the farmer. This rise in farm incomes was accompanied by a perceptible improvement in the economic conditions of the farmers, which was reflected in the assets acquired and the pattern of consumption resorted to.

This Table 4.15 represents the farm income earned from main products and byproducts. The farm income derived from main product is Rs. 22,93850 and the income derived from byproducts is Rs. 5,40800. The income derived from main product was higher than that of income derived from its byproducts.

The Table 4.16 assesses the income from other sources (production and quantity sold). The production of milk and milk products was worth Rs. 68,000; other farm products such as timber manure, etc. was Rs. 125300 sale of livestock and poultry was Rs. 2,30,000; Rent and Share was Rs. 10,000; income from fixed assets was 90,80950; rent from tractor/tiller was 25,00,000 and through others was Rs. 1,20,000.

The cost of selling was same for rent and share; income from fixed assets; rent from tractor/tiller and others. The mean for produced was 17,33464.29 and for sold is 17,23,278.57.

The Table 4.14 represents the monetary values of inventory and farm assets. Land and building were valued at Rs. 93750000 and it was highest compared to values of irrigation sources, agricultural implements and livestock. The total value of livestock as assets was Rs. 750000; they have Rs. 15000000 worth agricultural implements and Rs. 5250000 worth irrigation sources.

It was also found that, the income derived from main product was higher than that of income derived from it byproducts. In addition to agricultural activities like milk and milk products, other farms produce such as timber manure, sale of livestock and poultry, rent and share in co-operatives, income from fixed assets, rent from tractor/tiller and other farms products. The cost of selling was same for rent share, income from fixed assets, rent from tractor and tiller and other in case of sales. The tube well owners also have more assets after tube well irrigation. Number of assets like tractors, cycle, moped, bullock, and pumpsets has increased in selected villages and thus it brings about remarkable improvement in the living conditions of the farmers. Therefore, after the availability of tube well irrigation sources production and income from paddy cultivation has increased.

6. Household Goods

Electricity was available in the village even before the sinking of tube wells; so, farmers used some electric appliances even before they switched over to tube well irrigation. After switching over to tube well irrigation, many more farmers started using electric appliances, radios, fans, etc. This increase in the use of household goods can also be attributed to the improved economic conditions of the households after they had taken advantage of tube well irrigation. It was found that, the number of households using furniture has increased after the introduction of tube well irrigation. The use of two wheelers has become widespread, indicating the preference of many households of this mode of transport rather than of bullock carts or horses or walking.

7. Food Habits

The impact of improved economic conditions is immediately noticeable on the consumption pattern of the households, which were either underfed or fed on inferior cereals prior to the introduction of irrigation. It is natural that the underfed households should consume more. In such a situation, surpluses realized in the initial stages would be consumed with a view to reaching the minimum required consumption level. They would not be marketed at all. It is expected that, with the increase in their income, households, which consumed inferior cereals in the past, would switch over to the consumption of superior cereals and change their food habits. The number of households consuming rice has increased after the advent of tube well irrigation. It may be noted here that the increase in the number of rice-consuming households was larger in the case of large farmers than in the case of small farmers. One large size farm household totally discontinued the consumption of *ragi*, possible because of its improved economic position and the lower status symbol attached to ragi consumption.

8. Banking Habits

The banking habit is relatively common among the richer farmers, perhaps because of their better savings capacity, particularly after the introduction of tube well irrigation. The number of debtors too, increased after the introduction of this facility, mainly because loans were obtained for agricultural purposes. Loans from commercial banks and co-operative societies were used mostly for irrigation and agricultural purposes, while those from private indigenous moneylenders were used for consumption-cum-agricultural purposes. The introduction of tube well irrigation also developed contacts between farmers and money lending institutions.

9. Market Contacts

The interdependence of economic activities accentuates market contacts between rural and urban societies. Rural farmers buy agricultural inputs and sell their output in urban markets. The introduction of tube well irrigation increased these market contacts, for the new cropping pattern required HYV seeds, chemical fertilizers and pesticides, which are not needed for dry cultivation. Moreover, the introduction of tube well irrigation increased productivity and, therefore, the production of traditional crops went up over and above the home requirements. This surplus, too, had to be marketed, and led to increasing contacts with urban society.

10. Output-Processing Industries

In the first place, tube well irrigation results in increased production of paddy. This crop needs to be processed. Hence, enhanced production of this product leads to establishment and development of output processing industry like rice mill. This is the another impact of tube well irrigation on agriculture.

11. Input-Servicing Industries

Irrigated agriculture needs modern inputs like hybrid seeds, chemical fertilizers, pesticides, etc. and machinery like tractors with accessories, thrashers, harvesters, etc. in a substantial quantity. This requirement of modern inputs and machinery paves the way for the establishment and expansion of input servicing industries. Further, the growth of these industries would also enhance employment opportunities, of course, of skilled and semi-skilled workers. The Government, at least in initial years, will have to take initiative in providing modern inputs to ayacutdars. Private firms would also be supplementing their efforts in this sphere. Besides input-servicing industries, service and repair centers would also crop up to repair agricultural implements in command areas. Thus, tube well irrigation promotes the growth of input-servicing industries and service centers. Further, location of these industries is likely to have a profound effect on the eventual cropping pattern and production of a given command area.

Some rich farmers would also come forward to establish the output-processing industries. Here, they provide additional investment opportunities to them. This process would again contribute liberally to the process of asset formation of the area. Further, they enhance employment opportunities in an impressesive manner. Unskilled, semi-skilled and skilled labourers would employed in these industries in large numbers.

The impact, therefore, of tube well irrigation on the mode of disposal of the produce, its distribution, the increase in the national product and the augmentation of farmer-household's income is very significant. It has also made its own positive contribution to all these factors.

REFERENCES

1. T. Satpathy, Irrigation and Economic Development, Ashish Publishing House, New Delhi, 1984, pp. 1-2.
2. Hassan District At a Glance-2005-2006, District Statistical Office, Hassan, p. 22.
3. Agriculture Census 2001, Hassan District At a Glance-2005-2006, District Statistical Office, Hassan, p. 21.
4. Mysore Gazetteer, Vol. V. 1930 p. 1003.
5. Hassan District At a Glance-2005-2006, District Statistical Office, Hassan, p. 23.

5

FINDINGS, CONCLUSIONS AND SUGGESTIONS

This chapter presents the findings and conclusions of the study of the impact of tube well irrigation on agricultural production with special reference to paddy cultivation in Hassan district of Karnataka. It is also proposed to offer some suggestions and recommendations for the policy making in the development of tube well irrigation and thereby agriculture.

A BRIEF INFORMATION OF THE STUDY

This study was undertaken to show that Karnataka is fully dependent on ground water resources since it has fully harnessed the surface water resources. Among minor irrigation schemes, tube wells have acquired much importance since other sources like canals and tanks depend mostly on rainfall. Tube well irrigation has benefited the farmers very much, increased the standard of living and produced its impact on income and employment.

Hassan district is selected for the study since it has the only perennial river, Hemavathi. Due to failure of monsoon, tanks and canals are unable to supply water for cultivation. So tube wells are the only sources for supplying water for cultivation and so they are popular in the district. The study mainly analyses the growth and pattern of tube well irrigation, its impact on production, income and employment and also on socio-economic conditions of the farmers in selected villages of Channarayapatna taluk in Hassan district.

SUMMARY AND FINDINGS

The analysis is based on both primary data as well as secondary data, which reveal the trend and variation of tube well irrigation on agricultural production in Hassan district. Majority of the household have educational qualification from nil education to Master Degree and some have Diploma and ITI certificates also. It is registered 1 and 3 households respectively. The household with four members in the family was maximum. About 32 households had four members in their family. The one household with nine members in the family.

It is observed that the members dependent on agriculture were maximum in the household having four members in their family. It was least in the household having nine member in the family. Maximum number of households cultivated through tube well irrigation with owned tube wells. But the land leased in and leased out land is very least number of acres could be observed. The tube well irrigation is dominant when compared to well irrigation. Irrigated land and unirrigated land used for agricultural activities is having similar trend, which could be observed in both the cases. Irrigated land is used to the maximum for paddy cultivation, but unirrigated land used for growing other crops.

The material inputs of the agricultural activities were also invested through owned and purchases. In this context highest amount was spent in purchase of materials and lowest cost could be observed in case of owned materials. The number of indications (in Table 4.10), it could be observed in the case of fertilizer and pesticides registered only 260 and 800 respectively in case of materials owned. But whereas in the case of purchased materials it was found to be higher amount i.e. 54050 and 37980 when compared to owned materials.

Number of hired labour used for majority of the farm labour was more than family labour. The cost of animal labour and tractor labour indicates the cost of production in main products i.e. paddy cultivation. When compared, tractor labour spending is more than the animal labour. The total cost incurred for irrigation was more than any other activity. The total cost spend on irrigation is more when compared to other agricultural activities, it has spent only lesser amount as observed. The monetary values of inventory and farm assets, land and building values were highest when compared to irrigation sources.

The income derived from main product was higher than that of income derived from it byproducts. In addition to agricultural activities like milk and milk products, other farms produce such as timber manure, sale of livestock and poultry, rent and share in co-operatives, income from fixed assets, rent from tractor/tiller and other farms products. The cost of selling was same for rent share, income from fixed assets, rent from tractor and tiller and other in case of sales.

FINDINGS

The following are the findings of the study.

1. The present study highlights that majority of the households belong to small farmers category.
2. Education of farmers has helped in the efficient use of resources in the study area.
3. Small farmers still dominate in tube well irrigation and leasing of land was negligible.
4. It was found that unirrigated land is equal to that of irrigated land in the study area. Irrigated land has been utilised to the maximum extent for paddy cultivation. But unirrigated land is used for growing other crops.
5. It was found that under tube well irrigation production and income from paddy cultivation has increased.
6. After the introduction of tube well irrigation system, both human labour requirements and animal labour requirements has increased.
7. The farmers have accumulated more assets after switching over to tube well irrigation. Number of assets like tractors, cycle, moped, bullock, and pumpsets has increased in selected villages and thus it brings about remarkable improvement in the living conditions of the farmers.
8. Investment on agricultural inputs and other expenditure were found to be very high.
9. Tractor power not efficiently utilized and products from allied activities were found to be unsold.
10. Family labour was found to be under utilized.
11. Identification of tube well point is still a traditional one.

12. Erratic supply of electric power is affecting severely.

SUGGESTIONS

On the basis of the findings the following suggestions are made for further research and study.

1. Water Users Association in every village should be formed so that water resources can be utilised properly.
2. Watershed management should be given more importance and water resource management in case of tube wells should be enhanced.
3. The hired labour should be decreased and wherever family labour is available it can be efficiently utilized.
4. Organic farming has to be develcped for sustainability.
5. Use of chemical fertilizers may be decreased and encouragement should be given for earthworm manure and organic manure.
6. Water storage tanks have to be encouraged more, and more unirrigated land must be brought under irrigation.
7. More lands have to be leased in for better efficiency and livestock should be increased for better prosperity.
8. More concentration of byproducts has to be thought of and allied agricultural activities should be increased.
9. To encourage production, free supply of electric power may be sanctioned to the marginal and

small farmers and energisation of all pumpsets of tube well owners.

10. The Government is requested to increase the procurement price of paddy; then only paddy cultivation will not be affected and the shift from paddy cultivation to cash crops may be discouraged considerably.
11. It is suggested that the nationalized banks and co-operative banks to lend more credit for the digging of tube wells and the conditions of borrowing may be liberalized.
12. The road transport facilities may be provided to the remote villages, so as to help the farmers to bring their products to the market area and avoid the role of middlemen.
13. Farmers should be properly trained to adopt package practices and modern agricultural implements to reap the benefits quickly and continuously.
14. The location of tube well point is at present done on the traditional basis, which should be replaced by scientific method and to provide proper guidance to the farmer at the time of digging tube wells by irrigation and Agricultural Department.
15. There should be a regulatory methodology to see that there is a considerable distance between one tube well and the another.

FUTURE RESEARCH

The study of impact of minor irrigation particularly tube well irrigation in the Hassan district has paved the way for other researchers to undertake studies for future research such as comparison between tank and tube well irrigation; impact of tank irrigation; impact of canal irrigation and the cost and returns analysis of various minor irrigation sources. This study also helps the policy-making authorities to adopt suitable policy and execute it more effectively.

CONCLUSION

An attempt has been made in chapter 1 to state the background of the present study and review of literature. Besides, the scope of the study, objectives, study area, reference period and chapter scheme have also been presented in the first chapter.

In chapter 2, research methodology has been made. This methodology contains the data source, a profile of Hassan district, sample design, method of investigation, statistical technique, data analysis, and limitations and to define certain concepts used.

In chapter 3, the development of irrigation in India and Karnataka has been analysed. This has revealed that the development of irrigation has got a specific place in India's Five Year Plans. The irrigation potential created increased from 23 million hectares in 1950-51 to 98.10 million hectares at the end of 2005-06 and utilization increased from 22.6 million hectares to 84.70 million hectares during the corresponding period. In 1950-51, only 0.1 million hectares were irrigated by tube wells, but in 2001, 16 million hectares were served by tube well irrigation. Tube wells account for 28 per cent of total irrigated area.

In Karnataka, the expenditure on major, medium, and minor irrigation projects during the 2003-2004 was Rs. 3013.73 crores and 30.91 lakh hectares of irrigation potential has been created. There were 35133 tanks, 4,74,174 wells, 2,58,161 tube wells and 64 reservoirs in the state during 2001. According to 2003-2004 statistical data, in the state, out of 2384037 hectares of net area irrigated, 743383 hectares was irrigated by canals, 3913229 hectares by wells, 774117 hectares by tube wells, 147068 hectares by tanks, 116656 hectares by lift irrigation and the remaining 211484 hectares by other sources.

In chapter 4, Agro–economic profile of Hassan district and characteristics of selected sample units are presented. The intention here is to provide some background to the study area in order to understand the nature of its growth process through time. Besides, the impact of tube well irrigation on land utilization, farm implements and minor tools, inputs, cost of paddy cultivation, employment and farm income on the basis of primary data have also been presented in the chapter 4. In final chapter, summary, Findings and Suggestions have been analysed.

To sum up, tube well irrigation has made a drastic change in the sample villages. In the study area, there has been upward trend in the number of tube wells and a change in the pattern of tube wells. More and more tube wells have been energised. The cropping pattern has undergone a forcible change. The traditional crop, *ragi* has been replaced by paddy. There has also been a remarkable increase in the level of income of the farmers and in the level of employment of the households. The households have also witnessed socio-economic changes. The literacy level has gone up and the banking habits increased. In short, the life-style of the sample village folks has under gone drastic changes in recent years.

Needless to say, the findings and the suggestions made in this dissertation are with reference to the study area only and hence they should be treated as suggestive rather than definitive for framing policies at State or National level.

BIBLIOGRAPHY

Books

Carruthers, I. and Clark, C., *The Economics of Irrigation.* Liverpool University Press, England, 1981

Chow, V.T., *Handbook of Applied Hydrology*, 1964.

Coward, E. W., *Irrigation and Agricultural Development in Asia,* Cornell University Press, Ithaca, London, 1980.

Dhawan, B.D., *Development of Tube well Irrigation in India,* Agricole, New Delhi, 1982.

—, *Irrigation in India's Agricultural Development.* Commonwealth Publishers, New Delhi, 1994.

Dutta and Sundaram, K.P.M., *Indian Economy,* S. Chand and Company Pvt. Ltd., New Delhi, 2007.

Fukuda, H., *Irrigation in the World,* University Tokyo Press, Japan, 1981.

Garg, Satya Prakash, *Ground Water and Tube Wells,* Oxford and IBH Publishing Co., New Delhi, 1984.

Gustafson, C.D., *Basic Irrigation Principles,* California Agriculture 31; 263-266, 1978.

Jha, V.M., *Irrigation and Agricultural Development,* Deep and Deep Publication, New Delhi, 1994

Kallur, M.S., *Irrigation and Economic Development,* Chugh Publications, Allahabad, 1998.

Keller, J. and Blisner, R.D. *Sprinkler and Drip Irrigation,* Van Hostrand Reinhold, New York, 1990.

Kramer, P.J., *Water Relations of Plants*, Academic Press, Orlando, Florida, 1983.

Misra and Puri, *Indian Economy*, Himalaya Publishing House, Bombay, 2007.

Misra, Kanak Manjari, *Irrigation and Economic Development*. Ashish publishing House, New Delhi, 1990.

Narasaiah, M.L., *Dams and Irrigation*, Discovery Publishing House, New Delhi, 2005.

Pal, S.P., *Contribution of Irrigation to Agricultural Production and Productivity*, N.C.A.E.R., New Delhi, 1985.

Pant, Niranjan, *Community Tubewell and Agricultural Development*, Ashish Publishing House, New Delhi, 1985.

Rao, V.M., *Impact of Irrigation*, Himalaya Publishing House, Bombay, 1974.

Saini, G.R , *Farm Size, Resource-Use Efficiency and Income Distribution*, Allied Publishers Private Limited, New Delhi, 1979.

Sankara Reddi, G.H., and Yellamanda Reddy, T. *Efficient use of Irrigation Water*, Kalyani Publishers, New Delhi. 2002.

Satpathy, *Irrigation and Economic Development*, Ashish Publishing House, New Delhi, 1984, pp. 1-2.

Singh, K.K., *Farmers in the Management of Irrigation System*, Sterling Publishers Pvt. Ltd., Delhi, 1991.

Singh, Sahu, *Agricultural Problems in India*, Himalaya Publishing House, Mumbai, 1991.

Somashekar, Ne. Thi., *Karnataka Arthashastra*, Sumuka Publishing House, Bangalore, 2005.

Vaidyanathan, A., *Water Resource Management; Institutions and Irrigation Development in India*, Oxford University Press, New Delhi, 1999.

Varade, S. B., *Drip Irrigation*, WALMI, Aurangabad, 1988.

Venkat Reddy, M., *Dynamics of Irrigation Water Management*, Ashish Publishing House, New Delhi.

Verruijit, A., *Theory of Ground Water Flow,* The MacMillan Press Lt., London, 1982.

Willcok, Sir William, *Ancient Systems of Irrigation in Bengal,* BR Publishers, New Delhi, 1984.

Journals

Annual Plan 2003-04, Government of Karnataka.

Dhawan, B.D., "Indian Irrigation; An Assessment", *Economic and Political Weekly,* May 7, 1988.

—, "Questionable Conceptions and Simplistic Views about Irrigated Agriculture of India", *Indian Journal of Agricultural Economics,* Vol. XL, No. 1, Jan –March, P –11. 1985.

Dhawan, B.C., "Irrigation in Eight Plan; A Critique", *Economic and Political Weekly,* pp. A38-A44, 1993.

Economic Indicators, Government of Karnataka, 2004-2005.

Economic Survey 2005-2006, Irrigation Department Govt. of Karnataka.

Gopalkrishnan, R., "Efficient Water use in Irrigation Projects, *Yojana,* Feb, 16, pp. 22-28, 1980.

Government of India, A Reference Annual, New Delhi, 2005.

Hand Book of Karnataka, 2004-2005.

Karnataka at a Glance, 2004-2005.

Mundinamani, P.S., "Cropping Pattern in India", *Yojana,* May, 16-31, p. 20, 1991.

Niravari Kayakalpa, Government of Karnataka, 1996-97.

Randhwawa, N.S., "Agricultural Productivity Problems and Prospects", *Yojana,* Vol. 34, Aug. 15, p. 75, 1990.

Statistical Abstract, Government of Karnataka, 2000-2001, (Publishing 2006).

Vaidyanathan, A. "Irrigation and Agricultural Growth", *Indian Journal of Agricultural Economics,* Vol. 42, No. 4. Oct-Dec. pp. 503-526, 1987.

F

G

H

I

K

L

M

N

P

R

S

T

U

V

W